I0000515

Because You Can:
Your Cybersecurity Career

A simple and practical handbook
with actionable insights for success

By

Ali Khan, Arlene Worsley
and Gaurav Kumar

Book Cover Design by ebooklaunch.com

Copyright © 2021 Riszq Inc., 10185191 Canada Inc. and Cyber & Sapphire Inc. All rights reserved.

Published simultaneously in Canada.

No part of this book may be used or reproduced in any manner whatsoever without the prior written permission of the registered copyright holders, except in the case of brief quotations embodied in reviews.

Disclaimer of Warranties and Limitation of Liability: The opinions expressed in this book are those of the authors at the time of publication and do not represent the opinions of any entity whatsoever with which the authors have been, are now or will be affiliated. These opinions are subject to change and revision without notice at any time. The authors and copyright holders make no representations or warranties with respect to this book or its contents, and, to the fullest extent permitted by law, specifically disclaim any and all such warranties, expressed or implied, including but not limited to any warranties of accuracy, completeness, merchantability, absence of errors or fitness for any purpose. No warranty may be created or extended by sales representatives or written sales materials. The opinions contained in this book may not be suitable for your situation and do not guarantee any outcomes. You should consult with a professional where appropriate. The authors and copyright holders shall not be liable to any party for any actual or alleged loss, damage or disruption arising as a result of, or in connection with, this book or its contents, including but not limited to any errors or omissions in the book resulting from negligence or any other cause.

Library and Archives Canada:

Names: Ali Khan, Arlene Worsley and Gaurav Kumar, authors
Title: Because You Can: Your Cybersecurity Career/ Ali Khan, Arlene Worsley and Gaurav Kumar
Description: First edition 2021
Identifiers:
> ISBN 978-1-7774990-2-0 (Paperback) |
> ISBN 978-1-7774990-0-6 (Electronic Book)
Subjects: cybersecurity career-handbooks, self-help, manual, etc. | information technology-handbooks, self-help, manual, etc. | cybersecurity education, certifications, etc.
Library and Archives record available at www.collectionscanada.gc.ca

Dedication

To our families
who unceasingly gave us courage and who stood by us
through tireless hours in pursuit of a dream.

To our students and mentees
who gave us inspiration for this book.

To our supporters
who cheered us on with profound encouragement.

Contents

INTRODUCTION

Today, expert advice on a cybersecurity career is scattered and difficult to find in one place. Most of the information is disparate and available via expensive courses, invitation-only webinars, virtual conferences, podcasts, published articles, blogs or through in-passing conversations.

That is why we wanted to provide this handbook you can take with you and consume wherever you are. The information you need to make informed cybersecurity career decisions is centralized, suiting your lifestyle as a professional or student on the go.

You may be considering the next phase of your career, interested in understanding what cybersecurity is about or curious about whether committing to post-secondary or continuing education is worth it. At the end of the day, it is about your future, your time and your money-a true investment in yourself.

Whether you are in high school, an undergraduate or a seasoned professional, this handbook is designed to simplify your exploration of a career in cybersecurity with practicality top of mind.

We have designed this handbook in a question-and-answer format where you determine which questions are most pertinent to you now, next week or down the road. Very much like how you are in control of your career, you are also in control of how you read this book-front to back, relevant questions only and so on.

As educators and mentors, some of the questions we commonly hear from students include and are not limited to:

What is cybersecurity?

Is cybersecurity only for people with IT backgrounds?
Which cybersecurity role is right for me?
Which certifications should I pursue?
How do I land a job in cybersecurity?

After reading this handbook, we expect you to gain an understanding of the cybersecurity industry, career specializations you can choose from and how to pivot into your desired cybersecurity career. You will also be equipped with the knowledge to accelerate your career, stand out above the rest and shine while doing what you love.

Transitioning into cybersecurity has brought significant success into our lives, and we hope our handbook will help you join one of the hottest career options today.

CHAPTER 1: CYBERSECURITY - WHAT'S IN IT FOR ME?

Simply put, there has never been a better time to consider a career in cybersecurity.

Cybersecurity jobs have consistently topped numerous job board rankings based on sheer demand by employers and high compensation. Dubbed as a "career of the future," it promises to be a rewarding career choice for those who are new to the field and for those who are mid-career looking to make a transition.

From a bunch of teenagers ("phreakers") exploiting telephone lines for making long-distance calls to state-sponsored attacks exploiting nuclear plants, computer crime has come a long way, and so has the need to protect the information and the assets that store and process it.

By the 1990s, as organizations and individuals put more information online with rudimentary-to-no security, it was soon realized that computer crime could be a source of substantial financial gains. This generated much traction during the early 2000s where the internet pretty much became a widespread insecure network. Furthermore, the question around internal controls and governance was also asked after a prolonged period of corporate scandals (Enron and Worldcom) from 2000 to 2002; this was when the regulators set some rules and held organizations responsible for what they report to the market.

Congress passed the Sarbanes-Oxley (SOX) Act in 2002, which held directors and officers personally liable for the accuracy of financial statements and the protection of key financial applications and databases that generated those statements. Along with other regulatory compliance requirements, the SOX act pushed organizations to consider information security beyond just securing the perimeter. Organizations were also challenged further to strengthen their perimeter by way of structured access control, physical security, and increased awareness and training. As a result, IT security became a more defined career with roles and responsibilities spanning across all business lines.

As our society increasingly embraces technological advancements—such as self-driving cars, artificial intelligence, distributed ledgers and robotics—newer cybersecurity problems will need to be solved. In that light, the demand for well-rounded cybersecurity professionals will only continue to grow at an exponential rate.

The demand

- A 2018 report by (ISC)[2], the world's largest nonprofit association of certified cybersecurity professionals, announced a global workforce shortage of nearly 3 million cybersecurity professionals.[1]
- According to Cybersecurity Ventures, the cybersecurity unemployment rate dropped to 0% in 2016 and has remained close to that level ever since.
- According to Forbes, there will be as many as 3.5 million unfilled positions within the industry by 2021.[2]
- The 2019 *(ISC)[2] Cybersecurity Workforce Study: Women in Cybersecurity* found that women make up approximately 24% of the overall cybersecurity workforce.[3] In today's socio-economic environment, more and more organizations recognize the opportunity of this statistic and are making a dedicated effort to recruit and retain women in the field for increased gender equity.

As research companies release these data points each year, it is evident that the skills gap continues to rise. Additionally, the skills, knowledge and abilities of cybersecurity practitioners rapidly evolve to stay ahead of (or at least keep pace with) the adversary. According to the 2017 Global Information Security Workforce (GISW) study, two-thirds of the 20,000 respondents indicated their organizations lack the number of cybersecurity professionals needed for today's threat climate.

It is clear organizations need to broaden their search criteria when looking to hire individuals who can potentially solve current and future cybersecurity problems. Just because a candidate does not have years of specific experience, professional degrees or certificates should not mean that she or he would not be a fit. Organizations need to look beyond these binary indicators and consider a candidate's potential, including attitude toward learning, business acumen, analytical skills and more.

High compensation

- The Bureau of Labor Statistics states that the 2018 median pay for "Information Security Analysts" is around $100,000 USD per year. Note that this figure excludes consultant, manager and executive-level salaries.
- Chief Information Security Officers (CISOs) can make upwards of $275,000 USD and significantly more in large U.S. cities. CISOs are one of the highest-paid executive positions in the technology industry.

Top job rankings

- CNNMoney ranks "Information Assurance Analyst" as the #5 best job in America with a median pay of $98,000 USD. "Risk Management Directors," often a senior-level role in cybersecurity, was ranked #2, with a median salary of $131,000 USD.
- U.S. News ranks "Information Security Analyst" as the #4 best technology job with a median salary of $95,510 USD.
- Glassdoor ranks the role of "Security Engineer" as #17 of the best jobs in America with a median base salary of $102,000 USD.

High job growth rate

- According to the Bureau of Labor Statistics, the job growth rate for "Information Security Analysts" is projected at 28% from 2016 to 2026 as compared to the average growth rate for all general occupations at 7%; and
- "Information Security Analysts" are one of the only eight jobs out of 818 that are projected to grow "much faster than average," where the average pay is over $75,000 USD.

Explosive industry growth worldwide

- In 2004, the global cybersecurity market was worth $3.5 billion USD. In 2018, the global cybersecurity market grew to around $120 billion USD, a growth rate of over 3200% (or 32 times).
- According to a research report conducted by Global Market Insights, the cybersecurity market is expected to surpass $300 USD billion by 2024.

- Global cybersecurity spending is predicted to exceed $1 trillion USD cumulatively from 2017-2021, according to Cybersecurity Venture's 2019 Cybersecurity Market Report.

Evolving threat landscape

- The 2019 U.S. Worldwide Threat Assessment, led by Director of National Intelligence Daniel R. Coats, stated cybersecurity attacks were the number one threat to the U.S., affecting politics, the economy, the military and critical infrastructure.
- With emerging technologies such as self-driving cars, artificial intelligence, blockchain, advanced robotics and the internet of things, technology will play a more pivotal role in 'everyone's day-to-day life. These advancements will increase the need for cybersecurity as hackers can cause greater damage.

Real-world cybersecurity impacts

- Numerous Fortune 500 companies have been hacked:
  - Marriot – Up to 500 million accounts hacked from 2014 to 2018
  - Yahoo – All Yahoo accounts (3 billion) were compromised in 2013
- When Equifax was hacked in 2017, the credit score and identity theft protection provider immediately lost $3.5 billion USD in market valuation, resulting in the costliest breach in corporate history. Since the breach, Equifax has recently stated that the cybersecurity incident has cost more than $1.4 billion USD in cleanup costs.

- The largest robberies in recent history were attributed to hacking, with Coincheck losing $534 million USD and Mt. Gox losing $487 million USD in cryptocurrency.
- For a visual, take a look at this example illustrating data breaches over the past few years from *Information is Beautiful*. You may recognize some major brands. https://www.informationisbeautiful.net/visualizations/worlds-biggest-data-breaches-hacks

COVID-19 AND THE "NEW NORMAL"

Since the beginning of 2020, the push to digital has happened in a less-than-ideal manner, leading to some businesses learning the hard way. COVID-19 has had a significant impact on how the world operates and continues to define the "new normal" around us. Inherently when businesses are unprepared for uncertainties like a global pandemic and pressured to take a specific route, security often plays catch up. As we saw in multiple cases, as countries went into lockdown and homes became offices and schools, the adversary exploited these uncertain and confusing conditions.

There has been a significant change in the dynamics around security awareness and training, business resiliency and the "elastic workplace." According to the Canadian Centre for Cybersecurity, there are cases of COVID-19 tracking maps that infect devices with malware, COVID-19 phishing emails with malicious links or attachments as well as spoofed COVID-19 websites. Fraudsters are also phoning individuals to tell them that they have tested positive for COVID-19 and need to provide their banking information.[4]

So, what can cybersecurity career seekers anticipate as businesses slowly reopen and realign their investment strategies? According to Ruhi Ranjan, Managing Director for Accenture's Advanced Technology Centers India: "Against this backdrop, leaders face the urgency and complexity of reopening their businesses. To outmaneuver uncertainty, reopening also requires a program of reinvention. This presents an opportunity—and a need—for many companies to build the competencies they wish they'd invested in before including to be more digital, data-driven and in the cloud; to have more variable cost structures, agile operations and automation; to create stronger capabilities in e-commerce and security."[5] Securely adapting to our new normal will be critical as humanity dusts off and settles into this digitized future.

A career in cybersecurity promises to be one of the most rewarding careers. By scoring top marks in areas like compensation, demand and growth, cybersecurity jobs are a suitable contender for the best jobs in the world. You have made a great decision diving into this field!

Chapter 1 Key takeaways

- Due to sheer demand, cybersecurity jobs have emerged as an explosive career globally, positioning the field at the very top of numerous job rankings.

- Governments and companies are seizing exceptional growth opportunities for women in cybersecurity by paving the way in educational systems and the workforce where women are supported to expand and succeed in various cybersecurity roles across all tiers of an organization.

- The "new normal" post-COVID is here and cybersecurity will help reinvent how we learn, work and operate as a society.

Chapter 2: What is Cybersecurity?

Historical underpinnings

Securing the things we attach value to is as old as humanity itself. The ancient Egyptians are known to have used hieroglyphic disguise (now known as encryption) to protect critical information. The ancient Greeks and Spartans used the transposition cipher by moving around letters, using an instrument called the scytale.

With the advent of the first personal computer, automation started to appear within the workplace between the 1960s and 1970s. However, the first personal computers were predominantly standalone and mostly under one individual's control. There were very few security concerns back then. In the late 1960s, the Advanced Research Projects Agency Network (ARPANET) created the first interconnected network. The first four ARPANET nodes were connected using 50 kb/s circuits between the University of California in Los Angeles, the University of California in Santa Barbara, the Stanford Research Institute and the University of Utah.

These connectivity requirements gave rise to TCP/IP, which is the internet's backbone today. Welcome to the connected world! While this allowed easy synchronization between two endpoints, it also provided insecure points for possible exploitation. Connected computers inherently create a threat environment where the various attack vectors can exploit networks and assets to gain unauthorized access to data or overwhelm networks and compromise availability.

It is all about data

We have certainly come a long way during these 50 years of interconnectivity. Right now, we are living through an era characterized by significant technological disruption, and this is different. It impacts our personal lives (and private matters) as much as it affects our professional lives. We have come to a stage where technology influences everything we do. From online banking to vehicles (and everything in between), there has never been such an interconnected world of devices, data and available bandwidth.

The amount of data these interactions produce daily is simply mind-boggling. According to Forbes, we create around 2.5 quintillion bytes of data each day.[6] If that is not easy to wrap your head around, think about this: We have created more data in the past five years than we have over the entire history of the human race. We have virtually entered an era where data is the next currency. Every second of every day, more and more decisions are becoming data-driven, utilizing technology— such as the Hadoop ecosystem, data lakes, NoSQL databases—providing actionable insights in near-real-time.

The use of these technological advancements has brought about a renaissance in anything conceivable, from space exploration to e-commerce. Assets of value—from business trade secrets to patient information—are now stored digitally. Moreover, where there is value, there is a risk of loss or theft. There is always an adversary in the technology space trying to gain unauthorized access to this information, modifying it, undermining its utility or making it unavailable for use.

This is where you come in. As a cybersecurity practitioner, you will be tasked to stay ahead of the adversary to support the confidentiality, integrity and availability (CIA) of information and systems.

The first "worm"

Bob Thomas programmed what is widely accepted as the first computer worm, an experimental self-replicating program called the "creeper." The creeper infected DEC PDP-10 computers running the TENEX operating system. It gained access via the ARPANET and copied itself to the remote system where the message, "I'm the creeper, catch me if you can!" was displayed. The Reaper program deleted the creeper.[7] Since then, computer attacks have evolved into various types such as:

- Financial theft / Fraud: Stealing of funds from individuals.
- Hacktivism: The use of technology to promote a political agenda or a socially motivated cause.
- Nuisance: To create disruptions in the regular utilization of technology.
- Privacy breach: Unauthorized access and leakage of Personally Identifiable Information (PII).
- Extortion attack: Hold information hostage in return for financial gains (e.g., ransomware).
- Cyber espionage: The use of technology to gain illicit access to protected information, typically held by the government or state.

C.I.A. Triad and beyond

The C.I.A. (Confidentiality, Integrity, and Availability) Triad is widely accepted as the three pillars of information security. It drives various security principles of which translates into standard practices implemented by multiple technologies. However, it is a common belief that the C.I.A. Triad might have some limitations. For example, it might not address critical areas such as utility, possession and authenticity of the information. The concept of utility in cybersecurity is related to the usefulness of information; possession is about the physical state, and; control and authenticity are about validity. These are also essential tenets when thinking about cybersecurity.

Risk and business alignment

Will you ever drive a car fast (or at all) if you knew there were no brakes? That should be an easy "no." Cybersecurity is similar to brakes in a car; without it, businesses would not expand and evolve because of the implications of technology risks. In other words, cybersecurity enables organizations to compete and grow in today's increasingly data-centric world.

Security cannot be an afterthought and needs to be baked into various business functions. The cybersecurity facet of and associated people, processes, and technologies need to uplift the business vision, mission and strategy. To that end, appropriate oversight through defined governance processes must be established so that organizations can adopt a risk-informed approach to counter the evolving threat actor. Irrespective of the size of the business and the technology landscape, these foundational aspects of information security ensure resilient business operations and instill customer trust.

From firewall configurations to background checks for employees, all security-related tasks involved should address organizational risk. If security controls do not address risk, they must be re-evaluated, repurposed or retired. It is essential to understand that risk is not a stand-alone term; it is a function of asset criticality, threats, vulnerabilities, and likelihood and impact on organizational assets. For example, a vulnerability affecting a particular operating system version would not be a risk if there is no system with that operating system in an organization's infrastructure. It, therefore, becomes imperative to identify, evaluate, rate and prioritize risks using appropriate threat modeling techniques so that risks can be appropriately qualified or quantified.

There is an increased reliance on partners, vendors, and subsidiaries in today's highly connected business and cloud environments; this vastly increases the threat surface in which businesses operate. It is important to note that although there are numerous benefits with these partnerships, various risks are also inherited. It becomes exceedingly essential to utilize risk-based management imperatives to control the supply chain.

Cybersecurity and the evolving adversary

Since Bob Thomas' time, there has been a significant shift in what the adversary is after. It is predicted the next world war is cyber warfare. The days where a hacker's primary motivation was for bragging rights are now long gone. It is clear that the motivation behind the highly sophisticated attacks we see today is financial, and thanks to technologies like cryptocurrency, these transactions can be entirely anonymous.

It then becomes evident that cybersecurity professionals need to be highly adaptive and innovative to tackle today's adversary. One of the reasons why air travel is safe is due to monitoring and tracking. Everything is monitored in near-real-time in the cockpit of a plane, allowing the crew to respond quickly to minor issues before they escalate into major ones. This analogy helps illustrate why monitoring and correlating events across the technology stack prepares organizations to be well-positioned against today's cybersecurity threats.

With breaches commonplace, practicing security hygiene is a mandate. Much like when we don't take care of our teeth, cavities can happen. The same goes for when organizations practice poor security hygiene, breaches happen. With regular and proper brushing, flossing and dental hygiene, the frequency and impact of gum inflammation or cavities are reduced. Similarly, regular and proper security hygiene-such as vulnerability scanning, continuous monitoring, robust detection and response- can help reduce the frequency and the impact of cyber incidents. As cybersecurity professionals, you must be prepared for deliberate and accidental harms, both from outside the organization and within.

Setting up technical controls for the infrastructure by looking for known attack patterns and signatures is barely enough now-the adversary is way ahead of that. Today, the adversary consists primarily of highly motivated individuals with the appropriate financial backing, a specific plan and are highly skilled. Aptly called Advanced Persistent Threats (APTs), they usually target well-researched vulnerabilities that are sometimes not widely known or protected. These are known as "zero-day" attacks. To protect from such attacks, security professionals need to look for Indicators of Compromise (IOCs) within organizations by actively hunting for anomalies and containing them before those anomalies become critical security incidents. This activity is known as "threat hunting."

To make it difficult for the adversary to break in, investing in technology is vital. However, blinking lights and giant screens mean little if businesses do not invest in instilling a security awareness culture backed by appropriate governance. Cybersecurity is also about protecting an organization's brand, not just its information systems or data. It includes practicing the concept of "due care," meaning organizations must conduct their business with a level of judgment, prudence and the determination required to safeguard their employees and stakeholders. The absence of due care could mean negligence and eventually result in legal penalties for organizations; this translates to cybersecurity as a business-wide mandate and not just "stuff" that IT does.

Chapter 2 Key takeaways

- The advent of the Internet introduced connectivity, and connectivity introduced exploitable vulnerabilities now known as cyber threats.

- With heightened reliance on computers and the Internet for automation and innovation, greater intricacies create a broader attack surface from vulnerabilities in the supply chain, unprotected legacy systems and the inherent risks emanating from the remote workforce.

- Cybersecurity is no longer an option. It is a business imperative and is a part of the overall existential cost of an enterprise.

CHAPTER 3: WHAT ARE THE CYBERSECURITY DOMAINS?

With new advancements in technology, cybersecurity continues to adapt and mature over time. So, where do you begin to learn the required building blocks to establish your cybersecurity knowledge base? This chapter presents an encompassing overview of the key domains of cybersecurity.

There are various cybersecurity courses and bodies of knowledge readily available. However, the Common Body of Knowledge (CBK) of the Certified Information Systems Security Professional (CISSP) certification is perhaps one of the most comprehensive resources to establish a knowledge base. The International Information System Security Certification Consortium, or (ISC)², administers the CISSP certification.

The CISSP is also one of the most sought-after security certifications today, and rightfully so. The first credential was established in 1994 and it has stood the test of time. As of January 2021, there are 147,591 professionals who are CISSP certified worldwide.[8] (ISC)² has based the CISSP CBK on foundational security knowledge and, as such, divides it into eight domains, which are summarized in this chapter.

When referring to cybersecurity domains or capability areas, it is best to keep on par with foundational and trusted material. As technology adoption transforms around us, most globally accepted frameworks adopt and include knowledge areas (categories and controls) to align with the change. Because the CISSP is foundational covering all practice areas and is neutral to technology or geography, it is best, in our professional opinion, to base the cybersecurity knowledge areas on the eight domains included in the widely accepted (ISC)² CBK.

Domain 1: Security and Risk Management

Security and risk management is the first domain within the CISSP CBK. This domain addresses an extensive spectrum of risk management and information security topics, starting with the preliminaries of security principles such as the CIA Triad defined below and broad information about compliance and security governance.

The security and risk management domain gradually builds the understanding of security ethics, security governance principles, regulatory issues, security policies and security organizational structure. It also explains business continuity planning security principles such as recovery time objective, recovery point objective and business impact analysis. This domain further explores security risk management, at an enterprise level, including in-depth analysis and risk mitigation techniques.

Topics covered within the security and risk management domain include:

- **Confidentiality, Integrity and Availability:** Also known as the CIA Triad, these concepts are fundamental to cybersecurity and are interwoven into all aspects of the practice. Every security vulnerability or security control can be analyzed using one or more of these fundamental concepts. For example, a distributed denial-of-service (DDoS) attack on a web server is an attack on availability, whereas stealing personal data impacts confidentiality.

- **Legal, compliance and regulatory issues:** To establish and maintain an enterprise security program, it is essential for cybersecurity professionals to be familiar with legal compliance and regulatory issues. The security and risk management domain covers information security regulations, guidelines, laws and specifications. Learning effective methods for preserving information security laws and regulations can save an organization from legal matters such as federal fines. For example, Google was hit with a £44 million fine over ads in 2018 under the European Union's General Data Protection Regulation. The GDPR fined Google for lack of transparency and absence of valid consumer consent.[jjjj+9]

- **Security governance principles:** The security and risk management domain adequately covers six security governance guiding principles: strategy, responsibility, performance, conformance, acquisition and human behavior. These principles should define, direct, and support an organization's security structure to support and uplift business operations.

- **Professional ethics:** Professional ethics is a crucial element encouraging prudent professional behaviors. There cannot be laws and rules against every undesirable outcome; therefore, professional ethics play a vital role in establishing actions helping safeguard organizational information and human resource assets. Simply put, professional ethics comprise of corporate and personal values with which to conduct business honestly and with integrity. Therefore, most professional organizations have a code of ethics. For example, the (ISC)² requires its members to fully support its code of ethics as outlined in the preamble and the four canons. This domain covers the concepts of professional ethics cybersecurity professionals should be aware of to guide them to carry out their responsibilities righteously.

- **Disaster recovery and business continuity:** Disaster recovery and business continuity are two operational components that help organizations thrive when untoward events occur. While creating business continuity plans, cybersecurity professionals work directly with business lines to understand the relevant metrics around recovery point and time objectives; this ensures business requirements are baked into recovery priorities if an unplanned disruption occurs. Business continuity planning, disaster recovery and business impact analysis are also covered. These concepts explain how collaborative planning and preparedness support businesses to continue providing value even in disastrous circumstances.

- **Security guidelines, policies, standards, and procedures:** Clearly defined information security roles and responsibilities and management oversight play a significant role in determining an organization's overall security foundational structure. An organization's security policies represent the directional requirements to protect an organization from cybersecurity threats, data loss, and associated liabilities. After framing policies, creating relevant standards that uplift these policies is the logical next step.

While policies are higher-level and do not change with technology, standards are specific to technologies and should be built with that in mind. This domain also helps outline each document's purpose related to the importance of enforcing security responsibility and accountability. Some security policies have various guidelines, which are suggestions regarding how the implementation of policies should occur. To establish effective policies, a professional must understand the business strategy, security objectives and guiding principles within the security and risk management domain.

- **Security awareness and training**: It just takes one click from an untrained and unaware employee to trigger a breach. To that end, a well-structured and customized training and awareness program needs to be established, which can empower employees to make the right choice when it comes to social engineering attempts and to understand their role in securing enterprise assets and brand reputation.

- **Risk management:** Risk management, at an enterprise level, is a strategic initiative that builds a common vocabulary for the identification of the likelihood and impact of various threat scenarios to the most critical assets in an organization. The ultimate aim of risk management is to help bring the overall risk in an organization to an acceptable level; this involves accepting certain risks while mitigating, avoiding or transferring others. A well-orchestrated strategy aims at developing a scientific approach in identifying threat scenarios that impact each risk based on its likelihood and impact.

An effective risk management program is designed around widely accepted frameworks such as the National Institute of Standards and Technology (NIST). A well-structured risk management program also enables a cybersecurity professional to perform a coordinated threat-based analysis and prioritize risk treatment strategies to safeguard an organization against loss.

Domain 1 Key takeaways

- The premise of security risk management is to bring organizational residual risk-or risk that remains after security controls are in place-to an acceptable level.

- Risk is a function of the likelihood and impact of threats exploiting vulnerabilities on assets.
- Risk can be treated by applying security controls (mitigation), sharing it with another entity (transference), avoiding the process that gives rise to the risk (avoidance) or accepting the residual risk (acceptance).

DOMAIN 2: ASSET SECURITY

Assets can be anything that an organization attaches value to. Organizational assets can be an employee, equipment, facility, information or its partners. The first step toward protecting assets is to catalog and prioritize them according to their criticality; this ensures appropriate risk and business impact analyses can be performed. Appropriate asset management enables the organization to logically evaluate the importance of assets, secure those assets based on criticality and maintain the right level of protection over time. As evidenced by the recent pandemic, some organizations found their critical business assets and associated processes were not cataloged and prioritized appropriately, leading to the seizure or compromise in operational abilities.

A cybersecurity professional should aim to acquire the asset and data protection knowledge and skills no matter their specialization domain or vertical industry. To keep an organization's data protected against loss, a relevant understanding of techniques and tools is required to safeguard information falling into the wrong hands. Domain 2 guides cybersecurity specialists to understand asset and information classification, privacy protection, ownership concepts, data security control, appropriate retention and data handling requirements.

Asset security involves principles, standards, and structures to acquire, operate, monitor, and retain assets securely. The chosen security controls must be risk-aligned in their ability to maintain the availability, integrity, and confidentiality of a system. The asset security domain also addresses the requirements for appropriately collecting and handling information during an asset's entire life cycle.

The regulatory environment is continually evolving, and organizations find it challenging to stay ahead of these. The knowledge shared in this domain should allow the reader to logically evaluate how changing regulations, like privacy, underpin asset security to evaluate and implement appropriate data security controls.

The classification of supporting assets and information structures is one of the key learning outcomes within this domain. The asset security domain consists of various subtopics, one of which is privacy protection. This topic consists of concepts, such as data ownership, data remanence, limitations, data processors on storage and collection, and data retention.

Topics covered within the asset security domain include:

- **Information and asset classification:** Classifying information assets is the starting point to ensure critical assets can have commensurate security controls assigned. In simple terms, classifying assets is a method of organizing them into groups based on their criticality, confidentiality, sensitivity and overall value to the business. A business impact assessment (BIA), for instance, can be input into the classification of assets and data. To ensure critical assets and data are appropriately protected, each asset and data set must be classified adequately into tiers. Adequate data protection starts with classification and involves considerations across legal foundations. Appropriate asset classification leads to well-articulated rules around usage, storage, retention and disposal. Asset classification and security policies bolster each other and uplift the organization's information security mandates and overall risk posture. Asset classification should be formulated by the asset owner with appropriate consultation from technology, security and compliance personnel.

- **Ownership:** Data ownership is a critical role with responsibilities needed to govern and regulate data adequately. According to The Economist, "The world's most valuable resource is no longer oil, but data." Furthermore, according to IBM, "Every day, we create 2.5 quintillion bytes of data—so much that 90% of the data in the world today has been created in the last two years." These statements provide an idea about the ubiquity of data and the power organizations attach to it. Establishing a governance model with appropriate data ownership is critical at the outset. Not all of the data produced and collected is public. Data sets containing personally identifiable, sensitive information or intellectual property need to be protected. This domain establishes a connection between data protection and the importance of a data owner to identify, establish and maintain the proper controls.

- **Protect privacy:** A professional seeking to initiate his or her career in the field of cybersecurity should be ready to embrace the knowledge of privacy and associated data protection requirements. With the emergence of increasingly stringent regulations such as the General Data Protection Regulation (GDPR), the power balance has shifted somewhat in favor of consumers. Enterprises now are under ever-tighter scrutiny for protecting the personal data of their customers and employees. The evolving regulatory landscape means data life cycle controls and overall governance (including data traceability and lineage) have become complicated. Therefore, this warrants a cybersecurity professional to understand how protecting privacy and security go hand-in-hand.

- **Appropriate retention:** Every organization needs to abide by varying data retention periods. As such, data must be appropriately discarded according to the retention and distribution policies of the organization. However, there is no standard global agreement that formalizes the length of time an organization should retain data. For some organizations, there might be a legislative requirement to retain information for a stipulated period. In other cases, organizations need to make their own decisions on how long to keep data. The legal and regulatory requirements differ from region to region and organization to organization. Data owners should also ensure that the retained information is of appropriate quality and is readily available. To make this availability definite, an organization must consider underlying issues, such as taxonomy and normalization. The data relevant to third parties, business management or partnerships is significant for any organization.

- **Data security controls:** Recognizing weaknesses within a security system and implementing appropriate security controls is one of the most critical tasks a cybersecurity professional will be entrusted to support. Domain 2 covers the data security controls topic extensively. Data tailoring and scoping is an essential approach in determining where a security control application is needed. The criteria for choosing a security control and its implementation should also depend on whether the data is in motion, in use or at rest. A cybersecurity professional must implement preventative, corrective and detective controls competently and as relevant to the data's state and life cycle stage.

- **Data handling requirements:** As data traverses through an organization's business circulatory system, appropriate handling requirements are vital for cybersecurity practitioners to understand. These handling requirements include labeling, destroying and saving sensitive information. Domain 2 provides various execution options for secure information handling.

Domain 2 Key takeaways

- Asset classification is vital to ensuring security controls are commensurate with the criticality of each asset.
- Establishing data ownership is central to appropriate governance around critical assets.
- Data handling requirements vary with both the state and stage of data. As a cybersecurity professional, a thorough understanding of regulatory and internal compliance requirements is essential.

DOMAIN 3: SECURITY ARCHITECTURE AND ENGINEERING

Domain 3 covers architecture and engineering concepts specific to cybersecurity. It addresses the mechanisms of establishing secure information systems and relevant architectures. These requirements are crucial in defining and designing a system's functionality to protect an organization from untoward incidents, such as human error, malicious acts, natural disasters or hardware failure.

Focused on the content covered within this domain, a cybersecurity professional can efficiently establish secure design artifacts and also determine strategies to position security controls within a system. The designed security controls aim to optimize the quality attributes of the system's foundational attributes, such as confidentiality, integrity and availability. Security policies are also fundamental in ensuring the system is handled appropriately to avoid intrusions or data loss.

Domain 3 also explains the management and implementation of security engineering using secure design principles, such as fail-safe defaults, least access privilege, open design, complete mediation and separation of privileges. The design and development necessities rely upon an organization's security strategies and selecting countermeasures and controls that meet design requirements.

The security architecture and engineering domain also cover vulnerability management, such as server-side and client database security vulnerabilities, cloud security, distributed systems, industrial controls, cryptographic systems, mobile device web application vulnerabilities and embedded systems. The topic of physical security is also covered in detail. This domain helps the professional ensure architecture design principles are well understood (e.g., consideration of encryption system vulnerabilities that may lead to cryptographic attacks).

Topics covered within the security architecture and engineering domain:

- **Engineering systems and secure design principles:** A security system is a series of standardized processes employed to draft the policies and procedures based on which security system implementation occurs. Security engineers and security architects rely on various security implementation systems to build infrastructures with an optimized security strategy. The system is used to obtain blueprints for establishing information security programs that can be applied to diminish vulnerabilities and control risk.

- **Fundamental concepts of security models:** In Domain 3, broad knowledge regarding security models and associated concepts are provided to enlighten a professional in designing security systems. Understanding security models provides a solid foundational basis and guidelines to embed security-by-design concepts while designing security architectures. Usually, a security model constitutes analytical and technical ideas outlined based on system specifications. For example, military and government organizations prefer models built around confidentiality, while commercial businesses favor integrity-focused models. A CISSP candidate must be aware of the Bell-LaPadula model, Biba model, Clark-Wilson model, Noninterference model, Brewer and Nash model, Graham-Denning model and Harrison-Ruzzo-Ullman model.

- **Security capabilities of information systems:** In Domain 3 of the CISSP, a detailed explanation regarding the information system's security capabilities are provided, including thorough coverage of hardware-specific design considerations. Systems should be built to address memory protection, cryptographic modules, virtualization and data commingling, hardware security modules and smartcards.
- **Security architectures, designs and solution element vulnerabilities:** No information system is entirely immune to vulnerabilities and intrusions. However, an adequate system infrastructure can ensure that a security system is robust in limiting or lessening the impact of potentially exploitable vulnerabilities. As the tools for inducing vulnerabilities and threat actors become more robust, the exploitation of potential vulnerabilities can ultimately overhaul the complete system or provide a threat actor with unauthorized access to critical assets. Domain 3 of the CISSP includes detailed information regarding vulnerability types, security solutions, security designs and architecture.
- **Web-based systems vulnerabilities:** These vulnerabilities are a result of weakness and flaws in a web-based system. Vulnerabilities can arise due to various reasons such as misconfigured webserver, application design flaws or inappropriate sanitization of form inputs. In Domain 3 of CISSP, a comprehensive understanding regarding web-based system flaws and associated solutions are established.
- **Mobile systems vulnerabilities:** While most system intrusions and vulnerabilities enter a networking system of an organization, there are some cases in which threat actors aim for mobile devices of authoritative

professionals. Hence, adequate measures should be implemented to mitigate risk or malware and ensure mobile data safety against unauthorized access. This matter is covered in Domain 3 of CISSP in detail.

- **Security principles applied to facility and site designs:** Embedded systems are slowly attaining a considerable share in the technological market and various computational system devices. For Critical Infrastructure (CI) systems, as an example, fast responsiveness is required. Thus, a real-time embedded system proves to be efficient as these have a tremendous demand for robust security. On the other hand, the primary purpose of a CyberPhysical System (CPS) is to monitor the physical response process. Mainly, cyber experts monitor and control the physical process that is backed by multiple security devices. Protecting human life and providing safety is the most crucial facet.

- **Cryptography:** Cryptography is one of the most prevalent and effective ways to protect critical data while stored or traversed through communication lines. As computer processing power increases, attacks against cryptography using brute force or cryptanalysis become more achievable. In this domain, cryptography, cryptographic attacks, encryption methods and standards are covered. Cryptography can protect data integrity and maintain its confidentiality while also providing authentication and non-repudiation, which affords assurance and accountability. Familiarity with various aspects of cryptography is essential to maximize the protection levels in a security system.

Domain 3 Key takeaways

- Learning the concepts of secure design principles and underpinning security models provides an overarching understanding of why security-by-default system design offers a solid architectural foundation.
- The guiding principles of secure design also apply to securing sites and facilities.
- As inherent system vulnerabilities propagate and the compute power for exploitability increases, engineering concepts need to keep up. For example, quantum computers are around the corner and, once available, conventional cryptography will be obsolete. Keeping up with these developments is vital for a security professional's career advancement.

DOMAIN 4: COMMUNICATION AND NETWORK SECURITY

Domain 4 of the CISSP CBK covers communication and network security. This domain addresses the transmission methods, network architecture, control devices, security measures and transport protocols utilized to maintain the integrity, availability and confidentiality of information systems over public and private communication networks.

This domain also addresses network basics involving Internet Protocol (IP) addressing, network topologies, network segmentation, wireless networking, the Open Systems Interconnection (OSI) Model, Transmission Control Protocol/Internet Protocol (TCP/IP) suite as well as switching and routing.

The scope for network security and communication is increasing rapidly over time. Virtual conferencing, remote working and switching and routing are a few relevant examples requiring trustworthy, secure and stable communication mediums. Network eavesdropping and disrupted internet connections are the most popular cyberattacks that can cause considerable harm for both individuals and organizations.

As businesses faced challenges by the changing dynamics due to the COVID-19 pandemic in the workplace, it became evident that the conventional approach to "securing the perimeter" fell short. Organizations quickly realized the need for a highly scalable and flexible "elastic network" that considers endpoints and data as an organization's perimeter instead of the conventional gateways and firewalls. This required a change in what a security professional needs to understand regarding communication and network security.

Developing software and communication tools on top of an insecure network and communication mechanisms can be daunting for a system architect and a system user. Therefore, it is essential to understand comprehensive approaches through which data and information are communicated and interchanged.

The domain of communication and network security explains the fundamental viewpoints of security issues within a network channel. There are two underlying communication mechanisms: 1) authentication protocols and 2) protocol for secure communication. Candidates preparing for the CISSP exam can use these two essential directions as a practical approach toward this subject.

Topics covered within the communication and network security domain include:

- **Secure network architecture design**: The integrity and security of a network's communication system can only be adequately established if engineers considered the principles of standardized network design during a network infrastructure establishment process. Online operations with the Internet's help have made things easier for end-users. However, the Internet is also a gateway for threat actors to enter an organization's network.
- It is the prime responsibility of a network architect to ensure that the data is undecipherable and safely reaches the authorized receiving party. A network security professional must also be aware of the fail-safe implementation, weakest link security, the least privilege model and relevant state-of-the-art techniques and models to perform all of these activities. These aspects are covered in Domain 4 of the CISSP to add to a security specialist's abilities.

- **Secure network components:** For network security optimization, various techniques and tools may be implemented in a layered fashion. These predominantly consist of hardware, software and processes that enhance defenses against multimodal vulnerabilities to an organization's digital systems. Hardware and software consist of multiple layers that limit various vulnerabilities from damaging, spreading and penetrating through the networking systems.
- The four fundamental components ensuring network security are Network Access Control (NAC), firewall security, Intrusion Prevention System (IPS) and Security Information and Event Management (SIEM) systems. The information about these fundamental network components is provided in Domain 4, and a candidate preparing for the CISSP examination must be aware of secure network components and each respective implementation.
- **Secure communication channels:** A communication channel is a network type used to transfer information between two disparate infrastructures. With rapid technological advancements (such as Software-Defined Networks or SDNs), communication channels are becoming more efficient and flexible. For example, for secure and more flexible voice communications, unified communication solutions are increasingly used. These have become progressively popular as they enable seamless communication between various devices.
- However, this standard creates several possibilities for threat actors to misuse security deficiencies found within technologies, such as multimedia collaboration, remote access, voice and data communications and virtualized networks. Hence, security professionals must

be aware of secure communication protocols such as Transport Layer Security (TLS), Secure Socket Layer (SSL), Swipe IP Security Protocol (swIPe), Secure Electronic Transaction (SET), Password Authentication Protocol (PAP), Extensible Authentication Protocol (EAP) and more. This domain of the CISSP addresses ways to determine network vulnerabilities and the implementation of secure network protocols.

- **Network attacks:** A cybersecurity professional commonly performs various tasks with which he or she may not be familiar. Some attacks aim to steal critical data, while others overwhelm a network in a Distributed Denial-of-Service (DDoS) outbreak. A network security professional must know the adversary's various plausible modus operandi as it carries out network attacks. This knowledge will empower a professional and provide insights about how to recognize and contain network vulnerabilities. In Domain 4 of the CISSP CBK, ample knowledge about network security attacks and associated countermeasures is provided. Some countermeasures against a network attack include patch management, limiting or disabling unused network protocols and devices, utilizing redundant network configurations and establishing an acceptable activity baseline to gauge any anomalous activity.

Domain 4 Key takeaways

- Networks are the digital equivalent of the human nervous system, making interaction and information exchange between various facets possible. A cybersecurity professional should advocate a defense-in-depth approach to build network protection from varied and layered controls.

- Emerging technologies, like Software-Defined Networks (SDNs), have afforded flexibility and scalability. However, these have created new threat vectors, which a cybersecurity professional should consider in a secure network's design and operation.
- The threat surface has expanded with ubiquitous remote work. Therefore, a cybersecurity professional should consider controls that look beyond the traditional perimeter and focus on securing the endpoint while using secure communication channels.

Domain 5: Identity and Access Management

Controlling access to data and locations is central to the core pillars of information security-confidentiality, integrity and availability. In simple terms, Identity and Access Management (IAM) is a set of processes and technical controls allowing for the selective restriction of subjects to places or information resources. The ultimate aim of IAM is to ensure a system can efficiently control an environment so that unauthorized individuals are prevented from accessing confidential resources while allowing authorized individuals to carry out legitimate activities.

In an IT organization, IAM focuses on managing and defining an individual's access and role privileges as network users and situations during which a user is denied or granted specified privileges. Most organizations still face profound challenges in establishing and maintaining a good handle on access management as technologies evolve, and people are on the move.

At the core of an IAM system is a unique digital ID per subject (user or system). After establishing the unique digital identity, the digital ID must be modified, monitored and maintained as people continue to operate, move within or leave the organization. IAM systems help provide technologies and tools to an administrator to track a user's activities, establish reports on activities, enforce policies, and alter a user's role as required.

The IAM technologies include physical or electronic systems designed to control who (or what) has access to a physical or information resource. These generally contain provisioning software, password management tools, monitoring and reporting apps, identity repositories and security-policy enforcement applications. Systems for identity management are widely available for cloud-based and on-premise deployments, such as Microsoft Office 365.

The IAM domain covers the concept of managing and provisioning the access and identities utilized for information systems and human interaction with various elements of an information system. A threat actor's ultimate goal is to gain unauthorized access to a system to get to confidential data. Therefore, an information security professional inherently invests an ample amount of time in selecting, implementing and operating controls for maximum protection.

This domain addresses the identification, authentication and authorization concepts of systems, services and users. These include single and multi-factor authentication, identity management systems, session management, accountability, proofing and registration, credential management systems and federated identity management. The identity and access management domain also address cloud-based and on-premise identity as well as third-party services. Principles of rule-based, role-based, mandatory and discretionary access control are also addressed within this domain.

Topics covered within the identity and access management domain include:

- **Logical and physical access control:** Access controls are reinforced to determine the individuals who can or cannot access the resources within a computing environment. There are typically two access control types: logical and physical controls and a blend or hybrid of both. While physical controls regulate individual access to IT assets or location, logical controls restrict computer network connections, data and system files.

The integration of logical and physical controls combined in an organization's framework can help centralize control establishment and achieve a coordinated response in case of security vulnerabilities. This mechanism helps enhance an organization's capability to comply with the policies that enable them to obtain a holistic view of risk management operations. Domain 5 of the CISSP CBK covers physical and logical controls to provide professionals with a broad knowledge about how to optimize secure operations in a computing environment.

- **Identification and authentication of devices and people:** The identity life cycle comprises of three stages: 1) a user or an entity asserts who they are (identification); 2) proves the identity (authentication); 3) gains access to what is permitted (authorization). These concepts and how security is built around them are constantly evolving due to highly connected environments. A cybersecurity professional is often required to navigate these complexities, such as the pervasive interconnectivity of our personal and professional lives, also known as the Internet of Things (IoT).

An effective identity and authentication system for devices must provide maximum protection throughout various operational stages of a device against cyber threats. A security architect should operate to discover the correct balance and determine the accuracy provided by the authentication strategy operating in every category of IoT devices. These methods for identifying and authenticating devices and people are some of the core concepts discussed in this domain.

- **Identity as a third-party service:** Identity-as-a-Service (IDaaS) is an extensive range of primarily cloud-hosted offerings providing identity and access management in a popular "as-a-service" model. Mainly, IDaaS is a technological aspect within the identity and access management domain, focusing on providing and operating a user's identity within the cloud and hybrid environments. The primary responsibility of an IDaaS provider is to assist in determining the authorized users and restricting sensitive data from unauthorized access.

  Third-party identity services include procedures, such as authentication, identification, accountability and authorization; this helps in provisioning and configuring cloud-based servers at a significantly faster speed, which in turn gives customers an on-time and highly flexible service. These cloud-hosted services also allow organizations to reduce their on-premise footprint, drastically reducing the operational load.

  As evidenced by the latest SolarWinds cyberattack, a risk-based approach should be adopted while consuming any service that includes partnering with an extended ecosystem or service provider.

- **Authorization mechanisms:** This domain also provides extensive knowledge to a cybersecurity professional around the various options of providing authorization. There are different access control methods and models based on which an authorization mechanism might be devised. These include role-based and rule-based access control, as well as mandatory, discretionary and attribute-based access control. Based on the factors which determine a business requirement, any one (or a combination) of these access control options may be used.

- **Access and identity provisioning life cycle:** An organization must procure formal procedures and policies to manage user provisioning, account, revocation and review. Account management deletion and creation are also directly addressed in the access and identification life cycle. When an account is initially established, appropriate privileges are given. However, these privileges should also be modified during the life cycle of an account. A consistent review of the account ensures authorization levels are consistent with the changing dynamics of users and systems. When an account is no longer in use, it must be disabled immediately as a precautionary measure to limit unauthorized access. In Domain 5 of the CISSP, this topic is discussed in detail and frequently asked in the examination.

Domain 5 Key takeaways

- The foundational aspect of controlling physical and logical access to locations and systems is the basis of this domain. Identities should be created and maintained over time such that they restrict unauthorized access while providing timely and reliable access based on business need-to-know.
- Integrating a third-party service provider to augment an organization's access management capabilities might be an excellent option to leverage by providing cost-effective, flexible and scalable identity services.
- Managing identities and associated access levels can be challenging as employees move within and outside an organization. Therefore, appropriate governance mechanisms, including access reviews and reporting, should provide a practical solution to this problem.

DOMAIN 6: SECURITY ASSESSMENT AND TESTING

Each technology stack implemented in an organization requires custom, out-of-the-box or hybrid configurations enabling successful and secure operation of those technologies. To ensure those technologies are protected, organizations must conduct routine tests by way of security and risk assessments, audits and controls testing.

The security assessment and testing domain focuses on the evaluation of associated infrastructure and information assets. This is done by utilizing various testing techniques to determine and address the risk manifesting in the information assets. The risk may occur due to design flaws, architectural issues, vulnerabilities within software and hardware, coding errors, configuration errors and various other deficiencies. These deficiencies may influence the ability of an information system to perform its intended operation. The idea behind these assessments and testing activities is to ensure the security professional finds and addresses the flaws and misconfigurations before the adversary does.

An application's continuous validation vis-à-vis organizational information security processes, procedures, policies and standards are also covered in this domain. The significance of ensuring security controls are uniformly and continuously applied; business continuity and disaster recovery plans are evaluated; and intended disaster scenarios are accounted for also fall within this domain. It also covers the importance of implementing efficient practices for building an audit program for testing and validating security controls.

Topics covered within the security assessment and testing domain include:

- **Test and assessment strategies:** Risk management in cybersecurity revolve around managing, understanding, addressing, and monitoring cyber risks within an organization. Testing and assessment strategies play a crucial role in developing efficient data protection policies and bringing and managing overall risks to an acceptable level. Security assessments test a security posture of a system through verification of applied controls and their intended operations.

  The chief purpose of conducting a security assessment is to evaluate any flaws within a current state security infrastructure. The security testers and system owners must pay attention to the security necessities to determine which controls to assess. Nessus and Wireshark are some tools commonly used to conduct a security test. Depending on the system, these tests are conducted using both automated and manual testing methods.

- **Security process data:** A collection of security process data is essential for an organization to verify that security controls and their implementation are operating as intended. Security professionals should ensure that the security process data collection includes provisioning for management review, account management, risk and key performance indicators, awareness and training, disaster recovery, business continuity and backup verification. This domain of the CISSP CBK provides a comprehensive understanding and knowledge of each of the above aspects. The domain of security assessment and testing provides vital information regarding multiple security process data and testing that provides overall assurance to organizations.

- **The audit program and security control testing:** An increasing number of organizations are now shifting toward a risk-based audit approach. This approach is used to identify the emerging risk in a security system to assist a systems auditor in deciding whether to perform substantive testing or compliance testing. An IT auditor relies upon operational and internal controls and information regarding an organization's business environment in a risk-based approach. The decision regarding this risk assessment type can help in relating the control's cost-benefit analysis with the identified risk. The overall evidence-gathering process helps an IT auditor determine if the audited operation is adequate and well-controlled. The topic of the audit program and security control testing is covered in this domain to assist the candidates in optimizing a security system's efficiency.
- **Test outputs:** Security testing is an efficient way to predetermine exploitable vulnerabilities that might compromise a system. Upon identification of security vulnerabilities, a developer can efficiently adjust the system through mitigating or compensating controls. However, analyzing and testing outputs from time to time is necessary to evaluate the system's overall efficiency. The two ways of output testing are manual and automated. Manual testing is slow because it is labor-intensive and subjective but also has the advantage of human discretion. Automated testing utilizes specialized tools providing faster and more accurate results. Regardless of the technique used, test outputs should ascertain security control effectiveness and determine whether the system's availability, confidentiality and integrity can be appropriately maintained.

- **Security architecture vulnerabilities:** The immunity of a network system against cyber threat activities is not guaranteed. However, the implementation of solid security policies and robust security infrastructure can keep vulnerabilities at bay. Threat actors are becoming sophisticated with technological evolution. Hence, vulnerabilities can enter a system with ease if the security controls are not strategically placed. There are a variety of sources through which security vulnerabilities are initiated. Domain 6 of the CISSP CBK covers client-side vulnerabilities, server-based vulnerabilities, database-level vulnerabilities, large-scale parallel computing vulnerabilities and distributed system vulnerabilities.

Domain 6 Key takeaways

- Assessments and tests are central to ensuring the security controls an organization invests in and maintains over time continue to provide the intended protection. Through exposure to real-world threat scenarios as well as the tools and techniques an adversary may use, security professionals can get a step ahead.
- When developing a security assessment and audit strategy, security professionals must determine the specific scope in close coordination with management and business lines.
- Guided by industry or business requirements, organizations can provide their customers and regulators assurance about their business practices and governance by having a formalized security assessment or audit framework, thereby instilling trust.

DOMAIN 7: SECURITY OPERATIONS

Domain 7 of the CISSP CBK focuses on various operational activities directed toward protecting critical data within an organization. Topics such as recovery strategies, operational disaster recovery, incident response, physical and personnel safety and business continuity are covered.

An organization aiming to strengthen its security system utilizes various security operational models to protect its information system against intrusions. Furthermore, an organization must perform security vulnerability tests and penetration tests as part of an effective security operations strategy. The fundamental elements of security operations include disposal techniques, proper storage, configuration management, change management and system hardening.

This domain also covers forensic investigation, which focuses on handling and collecting evidence, reporting and documentation and investigative techniques, and steps to fulfill criminal, operational, regulatory, and civil obligations. It also discusses effective monitoring and logging requirements from devices such as Intrusion Prevention Systems (IPS), Intrusion Detection Systems (IDS), Data Loss Prevention (DLP) and Security Information and Event Monitoring (SIEM) systems. Preventative controls involving sandboxing, firewalls, and anti-malware software are also a part of this domain.

Topics covered within the security operations domain include:

- **Investigations requirements and support:** Numerous use cases warrant investigative and forensic activities. There might be instances of a malicious insider, a policy violation or an operational incident requiring an organization to research and trace back activities that led to these. With proper investigation and situational analysis, established security controls help to thwart or minimize the impact of such incidents in the future. During the investigation, security professionals must accumulate proper evidence, establish reports with practical recommendations and use comprehensive standardized techniques during the process. In the CISSP CBK, a candidate can learn appropriate investigative procedures and requirements per standards like NIST SP 800-61.

- **Monitoring and logging activities:** An organization can use technology to investigate audit logs and analyze any unauthorized activity that has taken place or attempted within an information system, such as monitoring and logging services. Effective implementation of logging and monitoring services can significantly help an organization track malicious activities and configuration alterations. A well-structured logging and monitoring service helps evaluate audit and system logs in a cost-effective and timely manner. Also, trend analysis provided by logging and monitoring service assists in protecting data confidentiality and determining potential improvements. In the CISSP CBK, this is covered in detail to polish a cybersecurity professional's skill to detect and respond to anomalous activities.

- **Foundational security operations concepts:** Comprehensive understanding of the security operations concepts are essential to provide a security specialist with skills to protect a system against vulnerabilities. The knowledge of security operations concepts-such as chain of custody, fault management, disk mirroring, clustering, backup strategies, Security Operations

Center (SOC) framework and so on-help structure strong security policies and integrate sound controls within an information system. Learning these concepts create a strong foundation for a cybersecurity professional to implement relevant security policies.

- **Incident management:** Incident management within information security is the process focused on the detection, analysis, response and amendment of the security complexities within an organization. The CISSP CBK establishes an understanding of operational management that reduces the possibility of unfortunate occurrences. Despite optimizing a system using effective security policies, exploitable vulnerabilities might still exist. Incident response plans include defined severity levels by asset and data classification, escalation process (call tree), incident handling procedures, identified metrics to measure the plan's effectiveness and more. Due to the inevitability of such incidents, the CISSP CBK addresses the requirements to utilize fully assembled and regimented methodologies to timely detect and respond to security incidents.

- **Preventative measures:** With various evolutions in the field of cybersecurity, vulnerabilities are ever-present with increasing attack surfaces. Effective preventative measures should be executed to address exploitable vulnerabilities within an information system. IPS, IDS, firewalls, honeypot strategy, anti-malware, 24-hour third-party support and sandboxing are a few highly significant preventive measures addressed within the CISSP CBK. Information about these preventative measures helps to implement adequate preventative measures to secure an information system.

- **Vulnerability and patch management:** Maintaining system, application and device currency is likely the most operationally intensive task in security operations. Vulnerability and patch management are two primary processes or tools to assist in system optimization to ensure known vulnerabilities are adequately assessed for risks and patched promptly. Through a vulnerability management program, sufficient security controls are established and enforced to scan and identify potential vulnerabilities within a system. A patch management program is a formalized approach to establish governance around timeliness and risk-based intervention to identify, test and remediate system vulnerabilities over time. In the CISSP CBK, these are discussed in detail.
- **Recovery strategies:** Despite establishing sound security policies and implementing adequate controls, an organization will likely still experience system outages and security incidents. In case of a declared security emergency such as a major data breach, an organization must execute recovery strategies to cope with damage adequately. Recovery strategies identify critical assets through Business Impact Assessments (BIAs) and prioritize recovery efforts to reduce the likelihood of catastrophic impact to the business. In the CISSP CBK, a candidate is provided with knowledge regarding recovery strategies and recovery planning to manage unfortunate events adequately.
- **Disaster recovery processes and plans:** Information security disaster recovery is a procedure that prepares an organization to bounce back from an untoward event, such as a natural disaster affecting physical facilities as well as paralyzing ransomware

affecting critical data sets. Therefore, an organization must have an effective Disaster Recovery Plan (DRP) to combat damaging scenarios and return to regular operations within pre-established thresholds. Establishing a comprehensive DRP, testing the DRP and communicating it to management enhances business resiliency. Concepts include response recovery point objective (RPO), recovery time objectives (RTO), allowable interruption window (AIW) and more. A thorough recovery plan is necessary to sustain profitability, safeguard an organization's reputation and reduce operational risk. As experienced during the current pandemic, a thorough business resiliency plan is not optional but a business mandate.

- **Personnel safety concerns:** Personnel security is a crucial aspect of any organization, related to security or not. Personnel safety and security should be at the core of an organization's security program. As we increasingly see the convergence of manufacturing facilities and industrial control systems with IT, personnel safety should be a part of all protection strategies.

Domain 7 Key takeaways

- Security operations are where the "rubber hits the road" for an overall cybersecurity strategy; this is where all policies and governance models are battle-tested. Risk-informed and business-aligned activities, together, form the overall security operations strategy.
- Security operations tasks can be daunting because of the very nature of the related activities. Appropriate use of automation and orchestration can ensure these activities provide value to protecting an organization's critical assets.

- Business resiliency is at the core of security operations. Proper recovery strategies define critical systems and inform recovery priorities, so business functions can operate during and timely recover from untoward incidents.

Domain 8: Software Development Security

A security practitioner needs to ensure security is baked into the software code's DNA. If security is tacked on late in the software development life cycle, flaws and bugs are discovered resulting in more cost and inefficiencies. A software developer can unintentionally or intentionally introduce security vulnerabilities within a system such as missing data encryption, Operating System (OS) command injection and missing authentication.

Many experts implement hardware and software controls to combat this issue to ensure an organization can manage system risk. Secure software development life cycle process is integral in ensuring overall security, as most business processes are becoming software driven. As adversary practices become more sophisticated, they increasingly exploit software flaws from development practices that are not foundationally as robust as the operating systems. A recent example is the SolarWinds incident involving the adversary inserting a vulnerability called "SUNBURST" in a version build of its Orion® Platform software development.[11]

This CISSP domain discusses maturity models, maintenance and operation, change management, software development methodologies and requirements a development team needs to follow to ensure secure development practices.

Furthermore, topics such as source code vulnerabilities and weaknesses, software development tools, the security of code repositories, peer code reviews, configuration management with regards to code development, logging and auditing relevant to change management, application programming interfaces security, software security-related risk mitigation and analysis and acquired software security impact are also covered.

Topics covered within the software development security domain include:

- **Software Development Life Cycle (SDLC) security:** The assurance process of secure SDLC involves code review, architecture analysis and application penetration testing of fundamental elements in the development effort. There can be various benefits of perusing the security development life cycle, such as increased software security within an organization and stakeholder's awareness regarding software security concerns. Moreover, resolution of issues and early detection due to secure SDLC leads to a more cost-effective promotion of software code to production environments.

  An organization can also reduce its overall intrinsic business uncertainties by adopting secure SDLC practices. Organizations might opt to leverage popular and secure SDLC models such as Microsoft Security Development Lifecycle (MS SDL) and NIST 800-64 to enhance their overall security posture. In this domain of the CISSP CBK, SDLC models and implementation are discussed in detail to improve an organization's ability to enforce secure SDLC models effectively.

- **Development environment security controls:** Security controls support a system's security mechanism and restrict a vulnerability from potential exploitation. The three types of security controls are preventative, detective and corrective. A security professional must have comprehensive knowledge regarding all these controls so their implementation can be precisely planned. Development environment controls are efficient in boosting a system's development and moderating the impact of vulnerabilities manifested in software.

Development and associated teams, such as configuration management teams, should ascertain an environment-wide security planning program and policies to establish a security system that effectively limits risk and fosters a secure life cycle. Developing adequate access controls will ensure enough protection against vulnerabilities and unauthorized modification. In this domain, multiple areas discuss development environment security control crucial from the CISSP CBK.

- **Software security efficiency:** Software security is a vital aspect of assuring a system's effectiveness against various uncertainties. The malicious attacks are becoming more potent, resulting in organizations investing more time and resources in acquiring secure and trusted software that withstands such malicious activities. However, developing an effective software security program consists of various phases, such as gathering requirements, designing, development, testing, deployment, and maintenance. Furthermore, if an organization wants to preserve data or crucial information against vulnerabilities, drafting explications based upon secure software development can provide adequate security. From the perspective of the CISSP, a candidate must have a broad knowledge regarding this topic.
- **Acquired software security influence:** It is crucial to determine the impact of newly implemented software on an organization's risk posture. The risk specialists within an organization require methods to evaluate the IT product's security and its influence on an organization's operation. Risk professionals should acknowledge diverse methodologies to analyze the authenticity of software in securing a system

adequately. In the CISSP CBK, a candidate learns about various methods such as binary analysis that is efficient in determining security flaws in software and internationally standardized methodologies such as the IEC/ISA-62443 and ISO/IEC 27034 series.

Domain 8 Key takeaways

- From e-commerce to automobiles, our world is increasingly driven by software. From software conceptualization to product promotion to production environments, there are various software life cycle stages (e.g., initiation, requirements, architecture, design) where security best practices should be embedded. The later in the life cycle security gets involved, the more expensive and time-consuming remediation activities become.

- Newer agile concepts, such as DevSecOps and Continuous Integration/Continuous Deployment (CI/CD), have become ubiquitous. These ensure security requirements that are a part of development activities are not afterthoughts or optional. These require developers to be security trained and inherently approach software development securely; this requires moving away from the traditional or waterfall approach to writing code.

- Secure software development life cycle requires harmony among various Information Technology Service Management (ITSM) capabilities, including change, release and problem management; this ensures appropriate governance and the ability to manage software risks at an enterprise level.

CHAPTER 4: WHICH CYBERSECURITY ROLE IS RIGHT FOR ME?

In this chapter, we walk you through the various rewarding career possibilities within cybersecurity. Cybersecurity has evolved to be an entire professional field requiring numerous roles and levels of expertise. These roles may apply to many business environments, including public and private corporations, not-for-profit organizations, consulting and defense.

There is often a misunderstanding in the industry, especially for aspiring professionals like you, that the field of cybersecurity is only for "highly technical people." This misnomer continues to cause a lot of misrepresentation within the profession. Cybersecurity, as a matter of fact, requires both business and technical acumen with a broad range of skill sets. These skill sets vary from strong leadership, management experience, technical, analytical, marketing, sales and data-driven expertise. More and more cybersecurity hiring managers and recruiters are expanding their search, looking for individuals passionate about the field who bring both business and technical bench strength. The following table divides the various cybersecurity roles into simple categories.

Table 1

Management roles	Technical roles	Non-technical roles	Technical and Non-technical roles
Chief Information Security Officer (CISO) Information Security Architect Information Security Manager Cybersecurity Program or Project Manager	Security Engineer Incident Responder or Incident Handler Penetration Tester, Ethical Hacker Computer Forensics Expert Malware Specialist Cyber Threat Hunter Security Analyst or Security Operator Information Systems Security Developer	Cyber Risk Management Professional Cyber Legal Advisor Privacy Officer Cybersecurity Sales and Marketing Cybersecurity Researcher	Security Assessor Cybersecurity Auditor Security Consultant Cybersecurity Instructor or Trainer

Next, we take a deeper dive into each of these roles, key responsibilities, successful characteristics, skill sets and references to potential salary averages. It is important to note these references are for information purposes only to give you an idea about compensation. Averages are used where good statistical data is found to provide representative information. Compensations may differ globally, and the information provided is for reference only.

Management roles

Chief Information Security Officer (CISO)

A Chief Information Security Officer (CISO) is regarded as one of the highest positions within an organization that an information security professional can attain. A CISO is a key strategic role that is liable to maintain and outline the organization's long-term security direction, goals and work plan to ensure the data and information are protected sufficiently in line with the organization's overall goals and objectives. It is reasonable to state that a CISO is where the "buck stops" when it comes to cybersecurity responsibility within an organization.

The CISO role was previously not widely prevalent. With the ubiquity of formal cybersecurity operating models and the changing regulatory landscape, CISOs now have more extensive and highly accountable roles within organizations. An organization's overall business success is catalyzed if its CISO effectively directs the operational, budgetary and key information management and data security tasks.

Key responsibilities of a CISO

Overall, the CISO is responsible for establishing and directing the information security program, including overseeing the implementation of projects and initiatives, ensuring various business lines can operate securely and within an organization's risk profile.

- Security operations, intelligence and cyber risk: A CISO should be able to champion and provide direction to counter any security issues by guiding the provision of logical solutions that can be implemented in a timely manner to control and protect the business environment. To that end, a CISO should keep up with any security threats that apply to the business environment. Additionally, a CISO should be able to provide guidance around understanding the conceivable security dangers that may emerge because of treachery and information coercion.

- Reduction in the likelihood of potential fraud: Protecting data from external threats is not enough. A CISO should be able to protect data from insider threats. If any such event occurs, the CISO should guide the organization to plan, prepare and execute recovery and resilience operations. A CISO should also ensure that appropriate investigation and forensics programs exist to help uncover and remediate any possibilities of fraud from activities like collusion.

- Secure architecture and design: A CISO is responsible for managing and protecting information resources. A CISO should guide the creation of a plan plotting the procurement, sorting, and utilizing the right technologies to deliver on the information security program.

- Limited access to restricted data: A CISO should determine the business requirements that warrant access to restricted data sets, thereby ensuring that only authorized users can access these. A CISO should also be well-versed with the regulatory requirements around access to restricted data and build a governance framework, including attestation and certification processes.

- Incident detection and response: Relevant detection and response activities to support timely identification and remediation of anomalous activities are central to the successful operation of a security program. In turn, this ensures customer trust and brand protection. To that end, a CISO should ensure relevant strategies are in place to develop, implement and operate an incident management program with the inclusion of relevant internal and external stakeholders.

- Administration: A CISO should be able to compose reserves whenever required and have full oversight of security personnel (internal and external to the organization) to ensure that all security-related tasks are running correctly.

Characteristics of a successful CISO

To obtain this very strategic security leadership position and prevail at it, the aspiring CISO should have the following characteristics:

- Meet core objectives: A CISO should be able to line up a viable strategy and vision in perceiving, managing, and recovering from any security-related issues. A CISO should be able to work under pressure to provide underlying core strategies, prioritize tasks and set up an operational view to catalyze data security and risk management. A CISO should continuously evaluate the security risks at every stage to deal with any problems beforehand and provide protection aligned with the organization's business strategy.

- Leadership qualities: To sustain this highly strategic leadership position, an aspiring CISO must have strong leadership qualities. The leadership role of a CISO is critical to set the "tone at the top" and sustain foundational security operating model. One person alone cannot carry out the massive responsibilities that a CISO is accountable for. Therefore, a CISO must ensure she or he leads from the front while building an effective team.

To make teamwork effective and ensure the teams contribute to business growth, the CISO must have an innate quest to promote an organization's security vision. Furthermore, a CISO should be able to motivate her or his employees effectively and train them to make essential risk management decisions independently thus ensuring any problem at the ground level can be sorted with relevant ease.

- Strong communication skills: A CISO represents and speaks on behalf of the information security management team. She or he is bound to collaborate with upper management, vendors and stakeholders; therefore, it is necessary to have strong communication skills so critical points can be put forth assertively. A CISO should be a brand ambassador for security and help target audiences understand the security vision. Communication is key to a successful business; therefore, acquiring impactful communication skills is required.
- Learning spirit: A successful CISO should have an urge to learn and stay abreast of the latest technological advancements continuously. As the technology landscape changes, so do the inherent risk profiles and control mechanisms. Therefore, a CISO should have strong analytical skills to understand security issues and be able to provide risk-aligned and business-relevant solutions.

Skill sets and education required for a CISO

An individual working in a specialized security domain learns more in terms of technical skills and business acumen as he or she climbs the command chain. A CISO must remain open to learning different specialized expertise and affirmations to do justice to her or his position, some of which are mentioned below:

- A CISO should know the concepts of secure network design, security architectures and IT strategies as well as each respective application.
- A CISO should ideally hold a professional education within IT security as well as leadership-focused security certifications such as the Certified Information System Security Professional (CISSP), Certified Information Security Management (CISM) or Certified Chief Information Security Officer (CCISO).
- Aside from security knowledge and expertise, a CISO should also be well-versed with program and project management best practices.

Average compensation range (USD)

Compensation ranges vary globally, especially when comparing developed versus developing nations. To provide a statistical estimate based on relatively available data, the average compensation of a CISO within the United States is $162,000 USD, with most CISOs earning between the range of $105,000 USD and $225,000 USD annually.[12]

INFORMATION SECURITY ARCHITECT

An information security architect is a cybersecurity professional who oversees framing, establishing and conserving an organization's technology environment by efficiently handling the security architectural design. She or he is a cybersecurity leader overseeing a team of security experts. An information security architect focuses on all the subtleties and ways threat actors can utilize to break into an organization's technology fabric, subsequently planning the security framework so the likelihood of such incidents is reduced.

In this profession, the information security architect is continually working to keep up with new technology and various means a threat actor can use to break into the organization's infrastructure. An information security architect bears the responsibility of becoming familiar with new exploitation strategies and approaches to battle the latest security threats. Numerous experts recommend that a capable information security architect be aware of the threat actor's psychology and anticipate a threat actor's upcoming strategies to break into an organization's systems.

Key responsibilities of an information security architect

An information security architect is a critical resource for an organization. She or he designs and implements the organization's security technologies to reduce the likelihood and impact of a security violation. The following are the general responsibilities of an information security architect:

- Be aware of the organization's IT footprint: As an information security architect, familiarity with the information and data security framework is essential. She or he should be able to comprehend the organization's security controls and relevant risk profiles to shield the company from any potential dangers while ensuring maintenance and upkeep of the security system to ensure maximum protection.

- Keep the current protection mechanisms relevant: As increasingly innovative technological advancements occur, threat actors and security dangers also advance in skill and nature. An information security architect should have the skills to restructure, improve, and upgrade the security system while implementing required improvements to prevent the systems from any security breaches adequately.
- Run security vulnerability tests: Keeping a check on the security systems is an ongoing activity; it is anything but a one-time process. Consequently, an information security architect is relied upon to run security vulnerability tests frequently to identify potential exploits, missing patches as well as remediate gaps. A security architect also champions external and internal offensive security testing (e.g., infrastructure penetration tests) to ensure the IT infrastructure's vulnerabilities are identified and remediated in a prioritized and timely manner.
- Reporting security violations: If an organization's infrastructure is vulnerable to security exploits, it is feasible for a hacker to penetrate security systems. In this case, an information security architect should be able to recognize such infringement, immediately report it and take the appropriate actions to contain a threat from obtaining further access.
- Develop, review, and maintain security policies and standards: An information security architect plays a pivotal role in defining security policies and standards. In most cases, he or she participates in developing

technical security standards and ensures correct enforcement. A security architect is the closest to implementations of technologies like firewalls and routers, for example, so she or he can provide the relevant technical expertise to protect the organization aptly.

- Developing timelines for security upgrades: It is clear that organizations have continuous tasks, and each time an organization undergoes a significant change, security frameworks should be updated accordingly. An information security architect should be able to create relevant timetables for upcoming changes and roadmap activities as well as redesign frameworks as applicable.

Characteristics of a successful information security architect

- Leadership and technical skills: As the nature of the job proclaims, an information security architect should possess sufficient technical skills to comprehend the security systems efficiently. She or he should also have leadership characteristics that will help motivate the team. Security supervision can be a daunting task and proficient group cooperation is fundamental.

- Complex risk management abilities: An information security architect should have broad information on complex and evolving threat vectors; this enables him or her to participate in relevant risk assessments by bringing in threat scenarios to consideration. This, in turn, helps create theories and logical threat models which aid in actualizing complex cyber and technical requirements to business language; this helps disparate parties comprehend potential risk within an organization's operating environment.

- Excellent written and verbal skills: It is essential to have relevant technical skills to build and support security infrastructure. It is also necessary to have soft skills to fortify an information security architect's capability. Therefore, possessing strong communication skills ensures productive collaboration with security teams, internal clients and external vendors.

- Attention to detail: For an individual working with continually evolving security threats and analyzing changes in the technology stack, focus is key. Paying close attention to even the minutest configuration items could mean an impending security incident can be tackled early on and remedial action can be mobilized well.

- Thorough knowledge about the vendor landscape: An information security architect works closely with disparate technologies and respective vendors. Working hand-in-glove with the vendor community and knowing the latest updates, fixes and patch releases are critical success criteria for a security architect. Additionally, up-to-date knowledge of inherent and upcoming technology risks and the adversary's modus operandi ensures the security architect has a good handle on infrastructure vulnerabilities.

- Immediate response and problem-solving skills: An information security architect should be quick in acting against security vulnerabilities. She or he should have excellent listening skills and adjust his or her aptitude, as indicated by the organization's vision. As a security leader, the information security architect should have the skills to collaboratively influence and coach their team when structuring and reframing the organization's technology security.

Skill sets and education required for an information security architect

- A bachelor's degree, preferably in IT, computer science or cybersecurity, is generally required. A master's degree focused on security or computer science may be required if applying for more leadership-focused security architect positions.

- An information security architect should have intimate knowledge about the various operating systems—Linux, Unix and Windows.

- An information security architect should be an expert in wireless, wired, network security, network architecture, security authentication, DNS, routing and proxy services.

- An information security architect should be able to re-frame, understand and implement new security advances according to an organization's requirements.

- An information security architect should be well-versed in implementing IT strategies and mitigating system networking risks.

Average compensation range (USD)

Compensation ranges vary globally, especially when comparing developed versus developing nations. To provide a statistical estimate based on relatively available data, the average compensation of an information security architect within the United States is $146,000 USD, with average compensation ranging between $85,000 USD and $189,500 USD.[13]

Information Security Manager

The role of information security manager has various iterations of a job title depending on the organization, including cybersecurity manager, IT security manager or enterprise security manager. An information security manager plays a pivotal leadership role within an organization. Usually reporting into the CISO, an information security manager is responsible for safeguarding an organization's security networking system against threat actors, security violations and malware in addition to overseeing the security team(s).

When an organization's security framework or significant data is exposed to threat activities, confidential and critical information may be lost. An information security manager supervises the Information Technology Security function, ensuring the security team(s) are well qualified and equipped to carry out their tasks within the organization.

A proficient information security manager is crucial as she or he can be pivotal in ensuring critical information is safeguarded according to an organization's risk profile. Experts have evaluated that an organization with a skillful information security manager, who is well aware of an organization's security posture, can add to its development by managing and reducing the likelihood of a security incident and uplift an organization's overall resiliency.

Key responsibilities of an information security manager

An information security manager is responsible for performing numerous critical tasks; these may vary from organization to organization. Some of the more common responsibilities are listed below.

- Security management of users and employees: An information security manager develops programs to manage users and employees from a security perspective. He or she needs to understand the security culture within an organization. An information security manager also ensures an organization has the preventative and detective capabilities to safeguard privileged user activity; this is where specific users have elevated access to sensitive information such as system and database administrators.

- Evaluate security Capital Expenditures (CAPEX) and Operational Expenditures (OPEX): An information security manager is responsible for carrying out budgeting for the organization's information security department.

- Identifying and staying ahead of vulnerabilities: An information security manager plays a significant role in limiting damage from security incidents. She or he continually analyzes an organization's security controls (e.g., firewall, security codes and antiviruses) to detect vulnerabilities that may emerge over time and formulate a threat and vulnerability management program to control those.

- Manage backup and recovery to bolster overall availability: An information security manager also has the mandate to ensure an organization's information assets are operationally available. An information security manager directs an organization's backup and recovery priorities per the enterprise Business Impact Assessment (BIA). In case of an incident (e.g., denial of service), she or he should guide their team to identify and contain it promptly and prioritize recovery efforts to minimize catastrophic impacts.

- Examine and shortlist new security technologies and services: As technologies evolve and various vendors attempt to stay ahead of the adversary, an information security manager should be able to decipher and keep up with this changing landscape. However, not all technology implementations provide value to an organization's technical footprint and risk profile. In coordination with relevant architects, it is up to the IT security manager to ensure the selected vendor and their associated products continue to deliver the perceived risk reduction and controls alignment.

- Communication bridge between the upper management and staff: An information security manager acts as a key liaison between security staff, business stakeholders and senior management. The information security manager should be able to translate the message to cater to various audiences. For example, using technical jargon for senior management and business stakeholders would not be a great idea.

Characteristics of a successful information security manager

- Effective management skills: An information security manager needs to manage employees and their progress; hence, she or he must have practical team management and coaching skills. This role also requires assertiveness and decision-making at all levels, which is necessary to complete tasks effectively.

- Efficient communication skills: Collaborating with IT and other business lines is one of the most critical skills for an information security manager; subsequently, communicating tasks is one of the role's requirements. Solid communication skills ensure employees working under the manager can understand the guidelines given to them and perform the assigned tasks effectively.

- Technology knowledge: An information security manager should be committed to learning the technology advancements that way an organization's framework can be upgraded and protected using effective methods to combat the evolving threat landscape.

- Leadership skills: Strong leadership abilities are required to inspire employees to perform at their maximum potential. A capable information security manager is aware of utilizing leadership capabilities to get tasks done with proficiency. Leadership is also about practical negotiation and relationship-building skills as the information security manager works with disparate business and leadership stakeholders.

- Analytical skills: An information security manager should have well-established analytical skills to determine whether the security systems need to be altered to keep questionable situations at bay.

- Adaptability: An information security manager is a crucial asset for an organization primarily to support new security challenges. In such circumstances, understanding the situation and adjusting to it swiftly assists with moderating dangers in a timely manner. Although effective strategies help avoid hazardous events, an organization's system can still be exposed to unknown risks and, hence, a successful information security manager must be highly adaptable.

Skill sets and education required for an information security manager

- Ideally, an information security manager should have a bachelor's degree in a relevant technical field and five years of experience within the practice.

- An information security manager should possess effective communication, project management and group collaboration skills.

- An information security manager should strive to achieve the Certified Information Security Management (CISM) or the Certified Information Systems Security Professional (CISSP) certification.

- She or he should clearly understand cloud risk methodologies, third-party auditing, operating systems and network designs.

- An information security manager should have strong leadership skills and project management skills required to manage budget and resource planning effectively.

Average compensation range (USD)

Compensation ranges vary globally, especially when comparing developed versus developing nations. To provide a statistical estimate based on relatively available data, the average compensation of an information security manager within the United States is $114,000 USD, with most information security managers earning between the range of $77,000 USD and $150,000 USD annually.[14]

CYBERSECURITY PROGRAM OR PROJECT MANAGER

Behind every successful project is a program manager and/or a project manager. According to the Project Management Institute (PMI), skilled project managers are in high demand with an estimated 2.2 million PM roles needing to be filled yearly through to 2027.[15] Relevant job titles in cybersecurity include IT Program Managers, IT Project Managers, cybersecurity program managers, cybersecurity project managers and portfolio managers.

The key difference between a program manager and a project manager is the size of the project, the number of projects managed by the role and the necessity of the role. A program manager is an advanced role while a project manager is an intermediate role and a program coordinator is a junior role. Generally, large projects with multiple workstreams will have a program manager who oversees a group of project managers representative of specialty areas.

For a large organization looking to mature its security posture over a three-year period as an ISO-certified organization for instance, a cyber defense program may be initiated to achieve the business objectives. This cyber defense program would require a program manager who supervises project managers specializing in various workstreams such as Industrial Control Systems (ICS) security, Identity and Access Management (IAM), Data Loss Prevention (DLP), application security, vulnerability management, cyber law, change management and more. Project managers leading these workstreams would manage a team of cybersecurity Subject Matter Experts (SME). Each project manager within the program would be instrumental in planning, executing, monitoring, measuring and reporting his or her team's project performance to the program manager. Program managers and project managers alike would share the responsibilities of resource planning, budget management, conflict resolution, client relationship management, and vendor management to name a few.

Today's successful cybersecurity program and project managers are highly skilled scrum masters who are knowledgeable about security management and IT service models (e.g., Information Security Management Systems [ISMS], Information Technology Infrastructure Library [ITIL, IT Service Management [ITSM]), enterprise and risk management frameworks, agile development methodologies and more.

Key responsibilities of a cybersecurity program or project manager

A cybersecurity program or project manager is responsible for performing numerous critical tasks that vary from organization to organization or client to client. Some of the more common responsibilities are listed below.

- Lead teams: A cybersecurity program or project manager guides a team in their work activities to meet project objectives, timelines and budgets. He or she may oversee a team of five to upward of 25+ members barring the size and scope of a project.

- Project risk management: Cybersecurity program and project managers are responsible for minimizing the level of project risk to an organization in alignment with the enterprise risk management framework and enterprise risk matrix. This includes assessing risks to the project, planning contingencies as well as tracking and remediating issues by way of a risk register that follows the principles of RAID-Risks, Actions, Issues and Decisions. Risk registers and RAID logs may be lean or comprehensive, barring the size of the project.

- Plan, schedule and monitor critical deliverables: A cybersecurity program or project manager develops a project plan and work schedule aligned with the cyber strategy the project is supporting. She or he uses tools to help streamline or automate processes of scheduling and monitoring critical security tasks. A successful program or project manager also forecasts potential project collisions to ensure there are contingency plans to ensure the health of the team and the health of the project.

- Budget and resource management: A common adage is time is money. Therefore, successful cybersecurity program and project managers aim to complete a project on time and under budget whenever possible; this includes the rigor of budget management and resource management. As an example, program managers may oversee long-term projects that are largely complex with budgets starting at $3 million. Project managers may oversee short and long-term projects with budgets ranging from $20,000 up to $3 million. Program and project managers may scale the size and scope of resources barring need. Tools help program and project managers streamline or automate the monitoring and reporting of this process.

- Vendor management: A cybersecurity program or project manager has some responsibilities within the supply chain management life cycle based on the needs of a project. Outsourcing vulnerability management and penetration testing, for instance, may be a business need and a project manager may collaborate with supply chain to select a vendor of choice, review contracts and service level agreements, monitor the performance of the vendor, manage the budget and more. It is common to see large projects requiring many vendors to support the different cybersecurity specialized work streams.

- Client relationship management: Above all, cybersecurity program and project managers must ensure the program or project meets the client's needs. A positive experience with a client coupled with successful delivery equates to trusting relationships, long-term opportunities and referrals to prospective businesses or business lines looking for similar delivery of cyber PM services.

Characteristics of a successful cybersecurity program or project manager

- A strategic business partner and change agent: A cybersecurity program and project manager with skilled strategic leadership is paramount to any business. Effectively aligning project plans with the overarching business strategy that is time-bound and consistently measurable helps companies to meet critical objectives. The recent pandemic surfaced complex business issues that required cybersecurity program and project managers to smartly and creatively solve security problems while also becoming agents of change helping make businesses more resilient.

- An accountable leader: A successful cybersecurity program and project manager is a leader who can motivate through transparency, accountability and trust. Some effective leadership soft skills may include an open-door policy, active listening, clear communication and inspiring team members to achieve great things individually and as a united front.

- Clear communicator: Communicating with all levels of an organization, from the executive suite to business lines and dedicated teams, require the ability to translate technical cybersecurity language into easy-to-understand information. Communication can make all the difference in the success or failure of a cybersecurity project.

- Knowledgeable about cybersecurity and project management: Business leaders and clients prefer cybersecurity program and project managers who have the right level of experience with managing complex projects and, where possible, are certified PMs. Knowledge and real-time execution of cybersecurity best practices, lean, agile or hybrid PM methodologies are in-demand blended skills businesses seek today.

- Adaptable: More often than not, cybersecurity projects rapidly change course due to unforeseen circumstances. Examples of unforeseen circumstances may include a sudden and massive budget cut within the cybersecurity program or a natural disaster that has halted time-sensitive security activities of an offshore vendor within a critical sprint cycle. As such, a successful cybersecurity program manager or project manager must quickly problem solve, mobilize contingency plans and adapt to dynamic environments-calm, cool and collected.

Skill sets and education required for a cybersecurity program or project manager

- Ideally, a cybersecurity program or project manager should have a bachelor's degree in any field (Computer Science is an asset) and up to five years of experience within the practice.

- A program or project manager should strive to achieve the PMI Program Management Professional (PgMP) or Project Management Professional (PMP) designation, a PMI agile certification such as the Disciplined Agile Scrum Master (DASM), and a cybersecurity certification such as the Certified Information Security Management (CISM) or the Certified Information Systems Security Professional (CISSP) certification.

- She or he should strive to gain a strong understanding of ISMS, ITIL and ITSM models, enterprise and risk management frameworks and agile development methodologies.

- A cybersecurity program or project manager should have strong non-technical soft skills such as leadership, communication, organization, prioritization, conflict resolution, problem solving and adaptability.

Average compensation range (USD)

Compensation ranges vary globally, especially when comparing developed versus developing nations. To provide a statistical estimate based on relatively available data, the average compensation of a cybersecurity project manager within the United States is $142,000 USD with most cybersecurity project managers earning between the range of $93,000 USD and $228,000 USD annually.[16]

TECHNICAL ROLES

Security Engineer

A security engineer is a technical professional within an organization responsible for maintaining and recommending the most optimal device and tool configurations from a security perspective. One of the most critical characteristics of a security engineer is maintaining an organization's reputation by limiting any harm to the confidentiality, integrity and availability of the data that an organization stores and processes. Security engineers should be able to identify and configure the proper integration requirements for tools and techniques.

Key responsibilities of a security engineer

- Benchmark security practices and standards: A security engineer is responsible for evaluating the organization's ongoing security design and architectural practices. If an organization lags in risk-aligned security practices, security engineers should report to the applicable management chain about the possible upgrades and implement them after the approval.

- Developing strategies to eradicate existing security issues: An organization can have flaws and inherent vulnerabilities within its existing security system. A security engineer should be aware of all the techniques to determine existing defects and apply their skills to minimize security issues.

- Outline the security requirements: Maintaining an organization's security system is an ongoing process. A defined budget is required to handle security systems within an organization. A security engineer should be able to prepare a precise outline of all the security requirements and communicate them as applicable, which aids in appropriate budget allocation.

- Keep up with technology to thwart security threats: Threat actors frequently come up with new threat strategies to exploit vulnerabilities; sometimes, it is before these vulnerabilities have widely accepted protection techniques. These are referred to as zero-day attacks or exploitation of zero-day vulnerabilities. A security engineer should keep up to date with such zero-day vulnerabilities and implement configurations and updates for security tools to thwart these emerging threats.

Characteristics of a successful security engineer

- Ability to effectively consume knowledge: A security engineer is continuously exposed to new challenges requiring flexibility. Comprehending the situation and learning new tactics to tackle the matter are signs of a successful security engineer. Learning is an ongoing process, especially when working in a position requiring the management of unique situations.

- Command of soft skills: In addition to having technical knowledge, strong soft skills (e.g., communication, decision-making, teamwork, time management) makes a security engineer influential. A security engineer should be able to author analytical reports and effectively communicate them to relevant stakeholders. A security engineer is also responsible for educating his or her peers about new practices and standards; therefore, having precise communication skills adds to a security engineer's ability. An innate creative and analytical mindset also helps thrive in this role.

- Attention to detail: The security engineer manages sensitive data by running tests to analyze the strength of a system, which requires attention to detail. It is essential to be meticulous about each intricacy of the system. Minor issues can be determined and solved at or near inception to prevent serious problems down the line.

- Problem-solving and agility: Setting up and maintaining an efficient security system requires creating, communicating and implementing a tightly aligned security plan that enables effective business operations and protects the brand. Therefore, a security engineer should be efficient and quick to pinpoint flaws and solve the issues.

Skill sets and education required for a security engineer

- A security engineer should have a bachelor's degree, preferably in cybersecurity or a computer science field.

- A security engineer should have a broad knowledge of networking, virtualization technologies, operating systems and programming languages.

- A master's degree in a computer information system or certifications such as GCIH, CISA, CEH or OSCP are recommended for senior security engineering positions.

Average compensation range (USD)

Compensation ranges vary globally, especially when comparing developed versus developing nations. To provide a statistical estimate based on relatively available data, the average compensation of a security engineer within the United States is $90,000 USD with most security engineers earning between the range of $61,000 USD and $132,000 USD annually.[17]

Incident Responder/Handler

An incident responder is a professional within cybersecurity who is at the helm of a cybersecurity incident throughout its lifecycle. An incident responder has solid technical skills and is quick in recommending actions to dangers that may emerge within the environment as the incident is unfolding. She or he is an expert in determining system penetration activities and security breaches.

As a cyber incident first responder, an incident handler should be well-trained in real-life crisis management and IT Service Management (ITSM); this ensures a quick evaluation of threats and minimizes the effect of such threats as effectively as possible. She or he should be sharp-witted and comprehend the security environment enough to limit peril rapidly. An incident responder should also carry out Root Cause Analysis (RCA) to prevent such circumstances from reoccurring.

Key responsibilities of an incident responder/handler

- Triaging and categorizing incidents: The first steps after an incident is detected are triage and analysis. These steps are performed to determine if an event is an incident and determine the priority and severity of the incident. Proper triaging can be the difference between a timely response to cyber incidents versus wasting time on false positives. An incident responder is at the forefront of mobilizing timely remedial actions, communication and insightful reporting.

- Analyzing and acknowledging system intrusions: An incident responder mainly works under stressful circumstances to recognize incidents and system vulnerabilities. She or he incorporates prevention methods such as risk analysis, threat detection and security auditing to re-secure a potentially compromised environment. Controlling the risk levels within an organization can be daunting; hence, proactively planning actions to respond to alarming situations is crucial. An incident responder utilizes system and network forensics, penetration tests and root cause analysis skills to respond to an incident.

- Working to educate staff about cybersecurity: To have a productive and collaborative environment that aligns employees with the organization's security vision, an incident responder needs to educate staff. Awareness within the team about security procedures forestalls risky circumstances.

- Risk reporting: When a risk is identified in an organizational security system, an incident responder should report it to relevant stakeholders right away; this will enable appropriate parties to rapidly map out solutions to navigate and remediate the situation at hand.

- Problem management and communication: An incident responder is an initial representative who determines system intrusions; therefore, he or she prepares in-depth reports about the incident. The incident responder can then determine the root cause and fix it promptly to constrain further unsafe circumstances.

Characteristics of a successful incident responder/handler

- Innovative: When working within a team, an incident responder should have diverse perspectives to critically analyze scenarios from various angles to recommend effective solutions. Whenever an organization encounters risky situations, an incident responder should provide agile solutions that are practical and rational.

- Think beyond boundaries: An incident responder is exposed to different threats and threat behavioral attributes. A one-size-fits-all approach does not work when troubleshooting various security threats. An incident responder should map the right solution to the problem while using his or her bench strength as a critical thinker to determine how to prevent these issues from reoccurring in the future.

- Complete involvement: An organization looks explicitly to invest in an incident responder who has a strong knowledge of the organization's environment.

- Threat intelligence: Having access to threat intelligence is fundamental to excel as an incident responder. His or her ability to integrate threat intelligence makes the process of discovering current system risks easier. Since time is of the essence, appropriate threat intelligence feeds help identify what industry peers have implemented, the lessons learned and the critical success factors.

- Information sharing and collaboration: Incident response activities inherently involve working with disparate teams in stressful environments. When multiple people are involved and work together to operate sensitive affairs of an organization, strong collaboration is mandatory. An incident responder should be able to deeply analyze the questionable situations and share the gathered information with his or her team so that suitable solutions are determined.

Skill sets and knowledge required for an incident responder/handler

- An incident responder should be familiar with all the current concepts of security software, hardware and IT security solutions.

- She or he should possess advanced experience with Microsoft and UNIX operating systems, how these work as well as knowledge of kernels, processes, threads, cryptography, program compilers, etc.

- An incident responder should have complete knowledge of programming language and scripting.

- She or he has strong communication and collaborative skills.

- An incident responder should work efficiently under pressure and be aware of how to prepare and present an analysis report.

Average compensation range (USD)

Compensation ranges vary globally, especially when comparing developed versus developing nations. To provide a statistical estimate based on relatively available data, the average compensation of an incident responder within the United States is $72,000 USD. Most incident responders earn between the range of $50,000 USD and $110,000 USD annually.[18]

PENETRATION TESTER/ETHICAL HACKER

Penetration testers, also referred to as ethical hackers, aim to preserve an organization's security configuration by running offensive tests against the organization's systems, just as a threat actor would. She or he also creates reports identifying weaknesses to improve an organization's overall security posture. A penetration tester simulates a cyberattack on an organization's environment to detect any exploitable threats before an adversary does.

Key responsibilities of a penetration tester

- Establish new tests to run across different systems: A penetration tester's fundamental responsibility is to examine systems against exploitable vulnerabilities in a controlled environment. To achieve this, a penetration tester uses similar tools that the adversary might. She or he should establish new penetration testing mechanisms to run across various system segments to identify potential dangers by simulating exploits.

- Help establish penetration testing scope: As with any offensive security testing, establishing the penetration test scope is critical. Otherwise, untoward outages and business disruptions may occur. A penetration tester helps define and formalize the scope of the test and clear rules of engagement. She or he also consults with the right stakeholders while limiting the scope to ensure proper participation can happen and knowledge is transferrable.

- Look for areas that require physical attention: A penetration tester also runs physical examinations to inspect physical deficiencies in systems, network devices, servers, and physical access management. She or he deploys methods to exploit discovered vulnerabilities and outline solutions to manage physical security issues such as vandalism, humidity, temperature, access card theft and natural disasters.

- Find vulnerabilities in proprietary applications and software products: A penetration tester uses different testing strategies to identify common and popular software weaknesses within proprietary applications. A typical method is performing system and network security audits to evaluate how efficiently a security system conforms to an organization's current security policies.

- Identify potential improvements to the security system: An organization enforces security policies outlining rules and directives to evaluate and utilize valuable IT resources. A penetration tester analyzes the current security policies and standards to assess effectiveness, suggest enhancements and improve the relevant security documents to align to strategy tightly.

- Keep up to date about new malware and threats: To effectively test a system for potentially exploitable vulnerabilities, a penetration tester should have proficient knowledge of recent threat vectors and malware. She or he should be able to think like a hacker to help identify new adversary tactics. Maintaining and keeping up to date with new and emerging threats, such as zero-day vulnerabilities, is a vital success trait of a penetration tester. Having a research mindset goes a long way.

- Establish reports to communicate the latest findings: After analyzing a target environment through testing and research, the penetration tester documents findings and actionable tasks with their judgment. They create precise security reports and discuss results with management to draw rational solutions. Penetration testers also conduct retests to verify and provide feedback after the implementation of new security policies.

Characteristics of a penetration tester

- Excellent comprehension of networking and operating systems concepts: Every organization has a distinct mechanism to safeguard its valuable resources. Hence, a skilled penetration tester examines an organization's security systems to have a good grasp of networking and operating systems. This knowledge enables rendering specific tests and eventually produce customized risk-treatment strategies.

- Staying current: Both application and infrastructure penetration testing requires an offensive security mindset. An offensive security professional should have an aptitude toward constant learning to stay abreast of the evolving threat landscape.

- Most penetration testers acquire a previous technical understanding of data administration and network development. However, the key to becoming an expert requires understanding the specifications of each application and network deployment within the organization. A penetration tester should be an expert in coding and configuration settings while researching current advancements as required.

- Meticulous and analytical: Every organization has varying infrastructures. Strategies and tools that may have worked efficiently for one project may not yield a similar outcome for another. A methodological approach leads to ethical problem-solving as every event occurring within a security system is distinct from another. An efficient penetration tester must think beyond the boundaries and never shy away from implementing varying strategies to resolve situations; she or he should consistently attempt to think like a threat actor to determine expected moves. A penetration tester must be detail-oriented as it enables accurately identifying minor fragments among configurations that can make a significant difference in limiting a threatening attempt.

- Think like a threat actor: A skilled penetration tester predicts the threat actor's expected moves to probe the implied dangers. He or she deeply examines each aspect, no matter how minute it might be. A penetration tester thinks analytically and creatively to fit missing pieces together.

- Social engineering skills: Tricking and taking advantage of untrained and unaware people makes for a vital tool in an adversary's arsenal. Penetration testers usually conduct investigative procedures to manipulate their targets to gather information. Hence, a penetration tester should have decent social engineering skills.

- Impactful written and oral communication skills: A penetration tester communicates at various levels within an organization and engages with managers to suggest the scope of potential solutions. A penetration tester is also responsible for establishing technical reports; therefore, strong writing and communication abilities are necessary.

Skill sets and education required for a penetration tester

- A penetration tester should preferably have a bachelor's degree in IT specializing in cybersecurity or computer science.

- She or he should strive to achieve certifications such as the CEH, GIAC Penetration Tester (GPEN), GIAC Exploit Researcher and Advanced Penetration Tester (GXPN) or the EC-Council Licensed Penetration Tester (LPT) are preferred and helps a penetration tester build and demonstrate credentials.

- A penetration tester should have ample knowledge of programming languages, networks, operating systems, computer hardware, software and network tools.

- She or he should build a roadmap to learn about Intrusion Detection Systems (IDS), Public Key Infrastructure (PKI), firewalls, etc.

- A penetration tester should have networking knowledge such as public vs. private IP, MAC addressing, DNS, routers, OSI model, DHCP, etc.

- She or he should have strong social engineering skills.

Average compensation range (USD)

Compensation ranges vary globally, especially when comparing developed versus developing nations. To provide a statistical estimate based on relatively available data, the average compensation of a penetration tester within the United States is $84,000 USD. Most penetration testers earn between the range of $57,000 USD and $136,000 USD annually.[19]

COMPUTER FORENSICS EXPERT

You may question why a computer forensics expert is classified as a cybersecurity career as, traditionally, computer forensics and cybersecurity were two distinct fields. A computer forensics expert is a trained cybersecurity professional who aims to assess an organization's IT assets to determine, recoup and evaluate a questionable occurrence while preserving evidence. These occurrences may be due to an unfolding security incident or part of an investigation.

A computer forensics expert investigates digital equipment such as computers, electronic documents and hard drives to submit evidence to an organization. Computer forensics is a relatively new profession; forensics experts collaborate with risk and information security experts to gather evidence from IT tools to speed up the investigation.

Most computer forensics experts are employees or external consultants at international organizations, private firms or law enforcement agencies. A computer forensics expert works intricately to trace any cybercrime activities within an organization's IT system and collect evidence while preserving the chain of custody throughout the process.

Key responsibilities of a computer forensics expert

A computer forensics expert is a valuable asset as she or he contributes significantly to internal and external investigations. In a few cases, the forensics expert also works with external law enforcement agencies to help with investigations. The following are the general responsibilities of a computer forensics expert:

- Recoup data from affected digital devices due to unfortunate events: When an organization encounters an adverse event, a computer forensics expert's extensive knowledge helps trace what transpired. In case of a cyberattack on digital assets, she or he recoups digital evidence from the IT devices. A computer forensics expert should be aware of all the feasible tactics that can contribute to the successful information recovery from infected IT devices.

- Assist in the investigation process: The investigation process after the occurrence of an unfortunate event is extremely sensitive for an organization. A computer forensics expert should gather evidence to simplify the process of investigation and conclude effective results. This must be done while preserving the chain of custody for the evidence to be permissible in the court of law.

- Uncover the tracks: An organization experiencing abnormal events should be able to reverse engineer the same, where possible. For an organization to realize the shortcomings help it design a robust system that can withstand attacks like ransomware, malware and other system intrusions; this is when a computer forensics expert plays a vital role. She or he analyzes the situation to extract the evidence and determine the intruder's modus operandi. The risk specialist can, in turn, use the collected evidence to draft additional protection against future incidents.

- Suggest compensating controls for security breaches: No matter how well-designed an IT system is, keeping all system vulnerabilities at bay is next to impossible. After a system violation has occurred, a computer forensics expert's extensive knowledge is needed to help apply elective safety measures immediately to minimize further harm. In some cases, implementing immediate mitigatory controls is not viable. As such, a computer forensics expert's opinion comes in handy for devising compensating controls while a more permanent solution is determined.

- Critically analyzing the collected evidence: A computer forensics expert should collect logical digital evidence to support the ongoing investigation. Familiarity with cyberattack types gives the computer forensics expert command over the evidence assortment process. She or he provides vital support in critically analyzing a situation's gravity based on the gathered evidence.

- Detailed analytical reporting on the recovered data: A computer forensics expert should prepare reports of the recouped data. The analysis report will help determine the integrity of the data (e.g., data modification) and the extent of the damage.

Characteristics of a successful computer forensics expert

- Technical aptitude: A computer forensics expert should be aware of the relevant technicalities required to regain data from digital devices in an unfavorable situation such as cybercrime. A computer forensics expert should be sharp and have an investigative and analytical mindset as tasks of this nature would warrant.

- Meticulous: The duty of a computer forensics expert generally involves extracting evidence to assist in an investigation. To efficiently gather logical proof, attention to detail is necessary. A detail-oriented computer forensics expert is in high demand as more and more firms require such niche services.

- Understanding of applicable law: A computer forensics expert is tasked to support the investigation process and, in some cases, even digital crimes. The field of computer forensics is closely related to criminology. The evidence a computer forensics expert gathers could help determine the facets of cybercrime. Strong knowledge about criminal investigation and related criminal law helps draw out firm conclusions.

- Analytical skills: A computer forensics expert should have extensive analytical skills to separate relevant evidence from data heaps. Having analytical thinking skills is mandatory if a professional is considering pursuing a career in computer forensics. She or he plays a critical role in assessing the information, observing an event's pattern, noticing red flags and contrasting current events with past events to establish trends and patterns. Understandably, this requires an extraordinary analytical aptitude.

- Willingness to accept unsettling situations: During a computer forensics expert's career, he or she might encounter disturbing circumstances that might leave a lasting impact. For example, a computer forensics expert might discover digital evidence relating to a hate crime or even child pornography. These situations become very challenging, and he or she should be ready to accept this harsh reality and fully support investigations to help eliminate such crimes from our societies.

- Knowledge of cybersecurity fundamentals: Despite computer forensics and cybersecurity as two distinct fields, there is increasing convergence. To excel in computer forensics and efficiently gather evidence, she or he should know about the cybersecurity domains. Having a solid understanding of cybersecurity concepts helps increase understanding the violator's viewpoint, which permits a computer forensics expert to help constrain such circumstances.

Skill sets and education required for a computer forensics expert

- She or he should preferably have a bachelor's degree in IT or a computer forensics field. New fields of study, such as cyber criminology, are also available as super-specializations.

- A computer forensics expert should aim to hold the GIAC Certified Forensic Computer Examiner (GCFE) certification.

- He or she should have strong knowledge of how computers work, memory, slack space, security systems, Microsoft and UNIX operating systems.

- A computer forensics expert should acquire technical skills such as handling evidence, implementing e-Discovery tools and applying cryptography principles.

Average Compensation Range (USD)

Compensation ranges vary globally, especially when comparing developed versus developing nations. To provide a statistical estimate based on relatively available data, the average compensation of a computer forensics expert within the United States is $73,000 USD. Most computer forensics experts earn between the range of $49,000 USD and $117,000 USD annually.[20]

MALWARE SPECIALIST

A malware specialist is a technical professional who determines, inspects and analyzes the nature of cyberattack vectors such as viruses, ransomware, trojan horses, worms, rootkits and more. A malware specialist collaborates with other information security specialists to intently observe previous attack patterns to decipher potential new ones, providing maximum protection of IT, OT and IoT devices against dangers.

Malware is specifically dangerous for an organization as the adversary increasingly uses creative methods to trick users for a successful attack. An organization can lose its valuable data during a malware attack on its systems and overwhelm IT resources, leading to a potentially catastrophic impact on the business.

A malware specialist with ample knowledge of the compiled and interpreted programming language can fix the systems damaged during malware attacks. A malware specialist should be familiar with the software system of an organization. She or he is well-versed in reverse engineering has the power to establish a defense system that can protect against the most aggressive types of threats.

Key responsibilities of a malware specialist

- Trend analysis and classification of malware: To limit the cyberattacks on an organization, it is critical to identify malware types used to violate the system. Through this characterization, a malware specialist correlates the current events with past events assisting in designing and executing fixes and controls to thwart future infringement. Identifying malware may also allow an organization to identify the attacker and the source of the threat.

- Observe and combat: A malware specialist is pivotal when an organization is experiencing a cyberattack through a malware attack vector. She or he needs to deeply examine the assault and observe the pattern to draw potential solutions. A malware specialist should explore and recognize different malware types and associated techniques to formulate the threat path.

- Tools and techniques: After an organization has experienced an attack, a malware specialist puts forth a complete attempt to observe the attack pattern and the kind of malware. It is sometimes not possible to manually recognize a particular type of malware and the ensuing threat. Therefore, a malware specialist should be well-versed in using specialized analysis programs and tools to observe and identify the attack pattern. Reverse engineering and dismantling the malicious code(s) become simpler when a malware specialist is aware of all the details.

- Staying current: Mostly, the incident response team works on the frontline to identify a cyberattack at an early stage. A malware specialist is needed during the early phase of a cyberattack to clarify the attack type. She or he is focal in detecting and controlling the risk. After an attack vector is identified and managed, a malware specialist also plays a significant role in the recovery process. For this to happen effectively, a malware specialist should be in lockstep with all response team activities and actively contribute to determining the associated methods to control an incident's frequency and impact.

- Contribute to policy, standards and procedures: A malware specialist works closely with other risk and IT professionals to ensure relevant revisions to policies, standards and guidelines. A malware specialist's input is critical in ensuring these documents provide information around addressing various exploitable vulnerabilities.

- Contribute to malware protection tools development: A malware specialist is technically well-versed about various malware types and how malware types might affect an organization and its systems. She or he should assist in developing or procuring protection tools to ensure the decision-making process involves understanding the various malware and potential impacts.

Characteristics of a successful malware specialist

- Suggest new strategies to thwart malware propagation: A malware specialist needs to continually endeavor to become acquainted with new malicious software as it evolves. She or he should focus on the latest methodologies a threat violator utilizes to break into organizations. A malware specialist should be an expert in reverse engineering to combat cyberattacks, ensuring inherent weaknesses are addressed.

- Malware research: A malware specialist champions the malware research process. A malware specialist collaborates with other risk specialists adding valuable feedback to enhance the research findings. When malware propagates the IT systems, various people, process and technology flaws might contribute to the level of impact to an organization. A malware specialist should know about these facets as critical inputs in defining relevant threat scenarios.

- Strive to learn: Because the field of cybersecurity evolves rapidly over a brief timeframe, a malware specialist must continually strive to learn and keep apprised of relevant trends, technology, threats and threat actors. She or he should have a strong awareness of new malware and strategies.

- An eye for detail: Attention to detail helps a malware specialist observe every aspect of a cyberattack, as the smallest minutia could mean an entirely different attack pattern and ensuing impact. This characteristic helps a malware specialist demystify the malware so possible solutions can be determined to keep future discomfort at bay.

Skill sets and education required for a malware specialist

- A malware specialist should preferably have a bachelor's degree in computer science or a relevant technical field. A computer engineering degree might be highly applicable.

- She or he should have a strong knowledge of the various operating systems and system design principles.

- A malware specialist should be able to re-establish data structures and unknown format files as well as have strong anti-debugging, dismantling and unpacking techniques.

- She or he should know programming and scripting languages such as Java, C++, Ruby, Python and Perl.

- A malware specialist should also be proficient in preparing technical reports.

Average compensation range (USD)

Compensation ranges vary globally, especially when comparing developed versus developing nations. To provide a statistical estimate based on relatively available data, the average compensation of a malware specialist within the United States is $93,000 USD. Most malware specialists earn between the range of $66,000 USD and $118,000 USD annually.[21]

Cyber Threat Hunter

A cyber threat hunter is a cybersecurity specialist who uses technical skills and expertise to identify Indicators of Compromise (IoCs) within an organization. Organizations actively seek to collaborate with a threat hunter who strives to identify and mitigate internal cyber threats by closely examining anomalies. It is the prime responsibility of a threat hunter to critically analyze an organization's overall internal environment to distinguish normal from malicious or harmful patterns, such as traffic flows or access attempts.

With the rapid advancement in cybersecurity threats, it is very likely that an organization already has a bad actor in its infrastructure. A threat hunter enables an organization to proactively look for malicious indicators and control those early on before becoming a full-blown incident.

A threat hunter usually uses advanced automated tools instead of the old detection methods to detect advanced threats. A threat hunter utilizes his or her creative and analytical skills to examine a system's expected behavior, build a hypothesis to construct the hunt, and look to prove or falsify the said hypothesis. To provide valuable services to an organization, a threat hunter should stay updated about various threat intelligence sources and cybersecurity research.

Key responsibilities of a threat hunter

- Decipher advanced threats: Advanced threats utilize creative methods through multiple channels to gather intelligence (reconnaissance), persistence (stealth) and laterally move to critical IT infrastructure areas. She or he proactively looks and correlates various indicators of concern and red flags. Due to the volume of data a threat hunter needs to assess, she or he should be well-versed in using advanced tools, such as SIEM and behavior analytics.

- Proactive and not reactive: By definition, threat hunting is a preemptive tactic. It becomes a highly strategic function that uses logic and trend analysis to report anomalies with actionable insights for course correction. A threat hunter works closely with other information security specialists to identify current vulnerabilities and suggest possible solutions to senior management, including the CISO.

- Establishing threat reports: A threat hunter also helps develop reports by including detailed analysis about identified "red flags" along with remediation recommendations. These reports should include relevant details with specific actions required from various teams. These reports are valuable for an organization to proficiently control the risks within an organization by outlining fit-for-purpose strategies to evade future vulnerabilities.

- Work with incident response teams: A threat hunter primarily works within a security operations team and takes a lead role in an incident response team by quickly detecting advanced threats and suggesting remedial actions.

- Intelligence analysis: A threat hunter carries out in-depth intelligence analysis to observe vulnerable activities. Analyzing intelligence allows a threat hunter to logically assess and inform decisions by framing feasible solutions to maintain a resilient IT infrastructure.

Characteristics of a successful threat hunter

- Driven by situational awareness and logic: Identifying abnormal activities within an organization requires a high degree of familiarity with an organization's IT infrastructure design and traffic and data flow patterns. A threat hunter can successfully discover abnormalities within a system if she or he has full awareness of relevant technology stack and data flows about the target organization. Within an organization, a system's normal operations vary from time to time; hence, a threat hunter should be aware of both constant and variable factors within a system, e.g., user profiles and patches. Understanding past incidents and subsequent solutions help a threat hunter to maintain proper and relevant use cases.

- Be a trusted partner: A threat hunter must establish a dependable relationship with his or her team to protect an organization's systems efficiently against attacks. Additionally, she or he should build a cordial, trusted relationship with other business leads. Building trusted relationships with decision-makers proves to help in remediating identified IoCs promptly. Conversely, a threat hunter also needs to rely on her or his peers to effectively work with various teams in determining root cause and problem management life cycles.

- Be resourceful and acquire TIP (Tool, Infrastructure and Personnel) skills: A threat hunter should have access to appropriate resources and know-how; these may include threat intelligence research sources, community and professional research, and information in cyber forums. The infrastructure available for testing to a threat hunter should support a range where she or he can experiment with detected malware and other anomalies within a protected environment for examination. A threat hunter should have sufficient experience to guide her or his teams in proficiently understanding and maintaining an organization's system against threats.

- Smart hunting: A threat hunter should be familiar with a wide range of security products such as network filters, firewalls, web filters, security analytics tools, Security Information and Event Management (SIEM) and Intrusion Detection Systems (IDS). To further increase visibility into an organization's systems, a threat hunter can amalgamate, correlate and contextualize anomalies from information obtained from security tools.

- Forward-thinking: A threat hunter must be able to anticipate the next move of a threat actor. This characteristic is highly beneficial to establish efficient strategies in safeguarding an organization's security system against harm.

Skill sets and education required for a threat hunter

- A threat hunter should preferably have a bachelor's degree in computer science with a cybersecurity specialization.

- She or he should strive to achieve cybersecurity certifications such as CEH and CISSP.

- A threat hunter should have in-depth knowledge of the concepts of networking and operating systems.

- She or he is well-versed in data analytics, malware analysis, data forensics and tools such as network filters, firewalls, web filters, security analytics tools,

Security Information and Event Management (SIEM) and Intrusion Detection Systems (IDS).

- A threat hunter has experience in implementing prevention systems and IT network-based attack strategies.

Average compensation range (USD)

Compensation ranges vary globally, especially when comparing developed versus developing nations. To provide a statistical estimate based on relatively available data, the average compensation of a cyber threat hunter within the United States is $76,000 USD. Most cyber threat hunters earn between $51,000 USD and $117,000 USD annually.[22]

Security Analyst

A security analyst (including security operator and threat intelligence analyst) is usually one of the positions through which a cybersecurity professional kick starts his or her career. A security analyst plays a crucial role in performing activities such as monitoring, alerting and reporting that help an organization's operating environment is well-maintained from a security perspective. A security analyst is likely the first point of human intervention in an incident's life cycle.

The most critical element of a security analyst's job is assessing security configurations and alerts while following the recommended procedures and playbooks for remediation. A security analyst may also provide inputs to ensure organizational security pieces of training are practical and relevant. She or he also suggests approaches to improve the overall security configurations within an organization and follow the communication channels to report security intrusions.

Key responsibilities of a security analyst

- Monitoring and detection: A security analyst keeps a close eye on IT-related activities within the organization. These activities are referred to as security event management or security information management, whereby a security analyst monitors external sources to analyze the threats that may apply to the organization. She or he usually performs the all-important "eyes-on-glass" function in a Security Operations Center (SOC) and follows established procedures to triage and categorize the incident.

- Alerting: Once an anomaly is detected, triaged and categorized, a security analyst should be quick in alerting about the incident. A security analyst should follow the established playbooks to alert and communicate the detected anomaly to relevant stakeholders.

- Evidence collection and accurate reporting: A security analyst should assist in the investigative process by gathering information regarding the intrusion activities, occurrences and generating accurate reports.

- Carry out security analysis: For a thorough incident investigation, a security analyst is relied upon to carry out a detailed examination of the alert and assist in confirming whether the alert is a false positive or a true positive. This information nurtures the incident management processes by improving the effectiveness and maturity of security operations.

Characteristics of a successful security analyst

- Knowledge of information security: With the cybersecurity field evolving rapidly, staying updated with the latest advances is essential. A security analyst should strive to keep up with the ever-changing times and acquire knowledge of the latest practices and techniques. She or he should endeavor to learn new approaches and use those to help build an effective security program.

- Analytical skills: Any individual working in the field of cybersecurity should have solid analytical aptitudes. A security analyst should precisely observe the IT system and determine any upcoming risks to tackle it proficiently.

- Strong communication skills: A security analyst collaborates with other teams within an organization to generate practical remediations in security incidents. Clear and concise communication skills are an absolute must to inform relevant parties about potentially concerning security events effectively. Proficient written and verbal communication skills are needed to accomplish these tasks satisfactorily.

- Ingenuity: A security analyst should be a forward thinker. It is essential to have creative skills to think ahead of time. A security analyst should envision future dangers and attempt to discover appropriate solutions to combat the adverse situation.

- Detail-oriented: Often, threats are difficult to recognize with stand-alone, single-sourced information. A security analyst should zoom out as required to gain a view of all aspects of a security incident to understand the bigger picture. She or he can help prevent loss by distinguishing threats as early as possible.

- Team player: A security analyst's role supports many security roles discussed in this chapter. For example, a computer forensics expert might rely upon a security analyst to run reports or provide system logs. Similarly, a malware specialist might call upon a security analyst to run traces of a malicious code or user activity.

Skill sets and education required for a security analyst

- A security analyst should strive to obtain a bachelor's degree, preferably in a computer-related field such as computer science or information technology.

- She or he should be well aware of security controls implementation, security tools functionality and niche skills such as penetration testing.

- A security analyst should aim to obtain certifications such as CEH, CySA+, GCIA, GCIH, etc.

- She or he should have strong analytical skills.

- A security analyst should have the ability to collaborate with IT professionals and business administrators effectively and easily translate technical jargon into business language.

Average compensation range (USD)

Compensation ranges vary globally, especially when comparing developed versus developing nations. To provide a statistical estimate based on relatively available data, the average compensation of an information security analyst within the United States is $98,000 USD. Most information security analysts earn between the range of $73,000 USD and $126,000 USD annually.[23]

INFORMATION SYSTEMS SECURITY DEVELOPER

An information systems security developer plays a critical role in protecting sensitive information, such as trade data, financial records, and Personal Identifiable Information (PII) or Personal Health Information (PHI). She or he implements critical security standards through the software engineering life cycle—from the assessment of prerequisites to the advancement and the application stage. The responsibility of an information systems security developer can fluctuate depending upon the nature of an organization. She or he plans, builds, runs tests and analyzes the information system security throughout its development stages.

An information systems security developer is responsible for making sufficient improvements by actualizing techniques to advance security within an information system. She or he aims to integrate security protocols with programs and software applications at an organization. An information systems security developer has a practical understanding of the design, assessment methods and technology applications to efficiently meet an organization's needs. She or he implements suitable upgrades and makes alterations to ensure software efficacy and security. An information systems security developer also collaborates with the software development team to develop suitable tools to adequately secure systems.

Key responsibilities of an information systems security developer

- Secure software development: An information systems security developer is responsible for intently examining different software development aspects to inspect if any area requires attention. He or she is an avid champion of ensuring information security is "baked in" to the software development life cycle instead of being an afterthought. An information systems security developer also aims to integrate efficient upgrades within the present systems and programs by implementing new security standards and policies.

- Comprehensive long-term software security strategy: An information systems security developer establishes methods applicable to protect systems holistically. He or she effectively comprehends complex technical requirements and maintains an information security system in a fast-paced business environment.

- Evaluate an organization's security needs: An information systems security developer should run tests on software components using penetration testing and incident management skills to adequately drive the needs of an organization's information security system. If required, an information systems security developer may onboard a penetration tester to conduct such tasks.

- Assist with security breaches: An information systems security developer helps analyze the information and networking system within an organization for potential security breaches; this is especially the case when the violation involves any custom code or piece of software. With the increasing use of Software-Defined Networks (SDNs), a system security developer is increasingly becoming involved in activities like threat hunting and breach investigation.

- Implement upgrades and protection: Information systems security developers utilize and install software, such as encryption programs and firewalls, to safeguard an organization's sensitive data. She or he also instructs employees about the installation or implementation of new security policies and procedures.

Characteristics of a successful information systems security developer

- Strong technical knowledge of programming languages: An information systems security developer should understand the life cycle of the software programs and software application stages proficiently. An information systems security developer should have adequate expertise in programming languages such as C++, Python and Java to decipher code vulnerabilities. She or he should be aware of SQL and other similar relational database languages as well.

- Strong communication skills: An information systems security developer can gain a lot of benefits through powerful verbal and written communication skills. She or he regularly interacts with the software development team to ensure the proper implementation of practical software upgrades. An information systems security developer communicates with clients to recognize and articulate security needs, frequently explaining and simplifying complex technical concepts and plans to non-technical professionals.

- Agility: An information systems security developer implements problem solving and critical skills at various software life cycle stages. When analyzing deficiencies within an organization's system, analytical skills help in evaluating and taking quick action. She or he should be agile in discovering fit-for-purpose solutions and employing security protocols in creative ways.

- Proficient in rewriting code and rebuilding protocols: An information systems security developer's main objective is to compose and implement software to restrict vulnerabilities from degrading the system; this can only be feasible if she or he can distinguish interventions and exposures. An information systems security developer should be able to rewrite code, implement various security tools and rebuild current security protocols. She or he should also operate efficiently under pressure and react effectively to security threats and indicators of concerns promptly.

Skill sets and education required for an information systems security developer

- An information systems security developer should preferably hold a bachelor's degree in computer science, software engineering or a related field.

- She or he should attempt to acquire security certifications such as the CISSP or CSSLP.

- An information systems security developer should have strong technical expertise in software development and knowledge of programming languages such as JAVA, C++ and Python.

Annual compensation range (USD)

Compensation ranges vary globally, especially when comparing developed versus developing nations. To provide a statistical estimate based on relatively available data, the average compensation of an information systems security developer within the United States is $98,000 USD. Most information systems security developers earn between the range of $66,000 USD and $143,000 USD annually.[24]

NON-TECHNICAL ROLES

Cyber Risk Management Professional

In the field of cybersecurity, no matter what your role involves, the ultimate aim is to control risk to a level that is acceptable to an organization. A risk management professional's role relevant in all areas within an organization, such as financial risk, marketing risk and operations risk. Specifically, a cybersecurity risk management professional helps keep an organization's IT assets functioning within its risk appetite. She or he develops and implements the appropriate risk management strategies at an enterprise level.

As outlined above, a cyber risk management professional works with senior management and business lines to define an organization's overall risk posture and ensure IT risks are managed within that threshold. She or he works closely with executives like the Chief Information Officer (CIO) and the Chief Information Security Officer (CISO) to effectively outline goals to manage risk and set up budgetary parameters. He or she analyzes the existing operations, standards and procedures for risk mitigation and communicates its efficacy to upper management. For instance, a risk management professional ensures that the organization is not buying a $10 lock to protect a $5 bike.

A cyber risk management professional also proposes distinct approaches to upgrade and enhance existing security measures and procedures. She or he establishes recuperation approaches and contingency plans when an organization suffers from operational availability incidents to efficiently recover impacted business processes within the Recovery Time Objective (RTO) and Recovery Point Objective (RPO) thresholds.

Key responsibilities of a cyber risk management professional

- Suggest strategies to manage risk: A cyber risk management professional performs various duties, all with a united purpose of managing risk within an organization's risk profile. The risk may appear in the form of an intrusion or data mishandling within the organization. He or she strives to identify potential risks and develop improved security measures to combat the expected uncertainties efficiently. If these risks are ignored or overlooked, some severe issues might ensue; hence, determining effective solutions is the key to minimize unfortunate events.

- Directing risk assessments to manage compliance: To control risks within the acceptable thresholds, a risk management specialist must continuously observe an organization's internal and external operations. He or she leads risk assessments to ensure compliance. A risk management professional should examine the potency of internal controls and check areas that are likely to present risk, and afterward take the initiative to refurbish those areas to diminish the impact and likelihood driving the risk.

- Document fundamental risks and communicate to senior management: After completing a risk assessment, a risk management professional is responsible for assembling information and data into streamlined reports, including statistics and graphs (e.g., risk heat map) to support final findings. The established reports are discussed with management to frame policies and plans to minimize losses and liabilities within an organization. A risk management professional also overlooks the enforcement and implementation of policies and standards.

- Manage the data quality of the risk management system: The quality of data plays a vital role in an organization's risk reporting dynamics. A risk management professional is responsible for managing the data quality in a risk management system to ensure appropriately informed decisions are incorporated to manage risks.

- Assist in implementing new technology: If an organization is looking to acquire new technology to enhance overall productivity, a risk management professional should help with the process to identify and communicate inherent risks. In turn, various stakeholders can make informed decisions about the acquisition and implementation of the new technology. Many organizations implement gating procedures to ensure appropriate risk assessments are a requirement at various stages of new technology implementation, and a cyber risk management professional champions this.

Characteristics of a successful risk management professional

- Quick to identify risks: As a risk management professional, one should have the ability to identify the various threat scenarios that apply to the business. A miss here can cause significant damage to the reputation and affect an organization adversely.

- Work under pressure: Some risks are unpredictable and can alter rapidly. A risk management professional must implement relevant strategies and take appropriate, timely actions in such situations. The established risk-based and business-aligned contingency plans and recovery approaches should come to the rescue in emergency circumstances.

- Negotiation: As a risk management professional, individuals must support business enablement through effective partnerships ensuring progression. Part of a risk management professional's job includes justifying the planned policies and upgrades to upper management based on effectiveness. Negotiation with other departments regarding risk scenarios, determining the appropriate security controls and obtaining staff buy-in to follow security standards are also required.

- Communication: A risk management professional should be clear and precise while communicating at any level within an organization; this ensures leaders and employees understand the current risk posture regarding security policy effectiveness and the changes required to provide consistent progress.

- Adapt to circumstances: Despite establishing effective strategies to address risk, uncertainties can still arise. These strategies also involve a combination of accepting, mitigating, avoiding and transferring certain risks. The objective is to bring the residual risk to a level that is acceptable to an organization. Nowadays, a risk management professional can utilize new technologies such as machine learning and leverage data analytics to manage upcoming risks. She or he should develop skills combining technology and knowledge to tackle uncertainties.

- Be proactive: A risk management professional should be proactive and continuously look for potential risk scenarios that may arise within an organization's information systems. He or she should be creative in establishing rational solutions to manage issues effectively and rapidly.

Skill sets and education required for a risk management professional

- A risk management professional should ideally have a bachelor's or master's degree in IT, computer science or any related field.

- To be successful in a senior risk management position, a minimum of five years' experience in a security role is highly desirable.

- She or he should possess a strong knowledge of information security procedures and systems.

- A risk management professional should strive to have excellent leadership, communication, problem-solving and interpersonal skills.

Average compensation range (USD)

Compensation ranges vary globally, especially when comparing developed versus developing nations. To provide a statistical estimate based on relatively available data, the average compensation of a risk management professional within the United States is $80,000 USD. Most risk management professionals earn between the range of $55,000 USD and $122,000 USD annually.[25]

Cyber Legal Advisor

A cyber legal advisor provides legal assessments and recommendations to an organization or clients on various important topics within cybersecurity, such as regulatory compliance, risk mitigation best practice, liability and insurance coverage, privacy protection, data handling, incident response and recovery, representation during an investigation and more. She or he advocates policy and legal adjustments. A cyber legal advisor establishes a case on a client's behalf through a wide range of verbal and written products, including legal procedures and briefs.

A cyber legal advisor evaluates guidelines and strategies efficiently. She or he defines how an organization may implement or abide by by-laws, determine clashes in policies and regulation, maintain familiarity with relevant constitutional matters, assess the consequence of adjustments to laws and guidelines, help execute laws and give legal guidelines organizations and clients. Often organizations require a cyber legal advisor to work on team projects to guide departments to identify adjustments and implement suitable data security measures.

Key responsibilities of a cyber legal advisor

- Provide legal advice to an organization or clients: A cyber legal advisor is vital in assisting clients or an organization with legal affairs. Hence, she or he should provide sound legal advice to help overcome and avoid liability. A cyber legal advisor also outlines laws or upgrades for implementation to effectively manage the organization's data security.

- Prepare legal documents: A cyber legal advisor leads and prepares required legal documentation on behalf of clients or organizations.

- Represent clients or organizations in court: If any legal restrictions are imposed on or by an organization or clients, a cyber legal advisor should be able to defend them in court. She or he should be able to formulate defense by collecting evidence or initiating legal action. A cyber legal advisor should adequately communicate with the client and witnesses to gather evidence for successful case proceedings.

- Analyze the case outcome: A cyber legal advisor should be able to efficiently determine the consequences of the case after each step, ensuring strong evidence gathering and powerful defense formulation to gain an advantage in the subsequent proceeding.

- Help implement data security measures: Usually, an organization asks a cyber legal advisor to assist in a project to determine effective data security measures related to regulations such as data residency. A cyber legal advisor probes the previous data security policies to identify any updates required to establish a relevant regulatory system.

- Examine legal data: A cyber legal advisor analyzes the legal data of clients or an organization to determine the soundness of prosecuting or defending lawsuits.

- Prepare for case presentation: In case of a legal allegation or lawsuit, a cyber legal advisor firmly leads preparation for the case. He or she should develop approaches, evaluate findings and establish strong arguments to gain beneficial outcomes.

- Monitor the impact of the latest technologies: A cyber legal advisor should analyze and monitor the potential influence of new technology on legal procedures and regulations.

Characteristics of a successful cyber legal advisor

- Effective communication: A cyber legal advisor should be proficient in presentation and interpersonal skills. Strong communication skills allow a cyber legal advisor to effectively state her or his point while defending a client or an organization in court for legal matters. This quality also proves useful when a cyber legal advisor collaborates with employees or potential witnesses within an organization to gather evidence.

- Agile problem solver: A cyber legal advisor should tackle various legal matters, whether proactively or during an incident or investigation. Each cybersecurity legal scenario occurring within an organization is highly diverse. Hence, a cyber legal advisor should be able to think critically and quickly on her or his feet to determine the logical strategies, recommendations or required evidence through practical problem solving to achieve the legal objective of clients or an organization.

- Approachability: A top-notch cyber legal advisor should have a diverse approach. He or she should analyze the situation from various perspectives when handling legal undertakings to structure rational solutions and provide substantial assistance in managing legal issues.

- Diplomacy: A cyber legal advisor should be diplomatic in representing clients or an organization in court. Diplomacy helps in effective negotiation between two opposing parties that positively influences the court's decision through dialogue. This characteristic holds true not only in the court of law but also in day-to-day business relationships and transactions.

- Conform to tight deadlines: A cyber legal advisor is continually managing various legal issues and lawsuits simultaneously. Hence, he or she should be able to manage workload efficiently and meet deadlines without fail.

Skill sets and education required for a cyber legal advisor

- A cyber legal advisor should have a bachelor's or a master's degree in a law-related field. There are many super specialization fields of study in the areas of IT and cyber law.

- He or she should have experience in a field related to the law when applying for more senior positions.

- A cyber legal advisor should be aware of the latest cybersecurity and IT technologies.

- He or she should know applicable privacy and cybersecurity laws and regulations to frame legal policies effectively.

- A cyber legal advisor should have strong communication, interpersonal and diplomacy skills.

Annual compensation range (USD)

Compensation ranges vary globally, especially when comparing developed versus developing nations. To provide a statistical estimate based on relatively available data, the average compensation of a cyber legal advisor within the United States is $80,000 USD, with an average salary ranging between $140,000 USD and $205,000 USD.[26]

PRIVACY OFFICER

As technology is becoming more prevalent, compliance has become a fundamental pillar for organizations and the way they do business. Organizations have prioritized compliance within their structure to ensure the security policies and procedures are considered through the exchange of any critical information internally or externally. As such, a privacy officer or privacy compliance manager plays a vital role in framing laws regarding information policies and procedures to mitigate risk and protect the Personal Identifiable Information (PII) and Personal Health Information (PHI) that an organization may hold.

A privacy officer or a privacy compliance officer is responsible for enforcing policies and supervising the application of data protection strategies to ensure synchronization according to the data protection laws and regulations per applicable jurisdiction. A privacy officer ensures the development, application and enhancement of efforts to adhere to privacy standards that provide safe data handling within an organization.

Key responsibilities of a privacy officer

- Design and improve information protection policies: A privacy officer should make adequate contributions to design comprehensive and strategic privacy policies and related procedures. She or he should outline, establish, sustain and implement privacy practices that effectively minimize risk and ensure the confidentiality of data when exchanged through a variety of mediums such as electronic or paper. A privacy officer ensures the existing policies are adequate in data protection and work to establish effective upgrades in case of deficiencies.

- Handle internal policies: A privacy officer should participate in risk assessments to help examine the legitimacy of established privacy policies and monitor the compliance-related issues within an organization. She or he takes the initiative to maintain internal procedures to ensure every individual in an organization is in sync with the data protection goals.

- Raise awareness among employees: It is the prime responsibility of a privacy officer to help with training to ensure data protection awareness among employees, especially those involved in information exchange. A privacy officer assumes a lead role in assisting employees and third parties in complying with the organizational privacy policies and procedures.

- Report any failures: A privacy officer participates in establishing and administering procedures that assist in investigating policy breaches and outlines immediate actions on security complaints. She or he also participate in security breach tests to determine, report and minimize risk within an organization. A privacy officer also works with the human resource department to ensure privacy violations are reported immediately so prompt responses can be taken.

- Investigates inappropriate access: A privacy officer works with the Chief Information Security Officer (CISO) to establish a system to help analyze, track and report policy violations. She or he observes patterns such as inappropriate access or policy violations to help formulate required upgrades to existing security controls or policies and thus further increasing the overall effectiveness of an organization's data protection system.

- Represents an organization's privacy interest: A privacy officer works with an organization's legal counsel, administration and other relevant internal groups to uplift the data protection interest to external stakeholders, such as customers and local or state governments.

Characteristics of a successful privacy officer

- Knowledgeable of IT risks: A privacy officer should comprehend threats associated with different levels of information sharing within an organization. She or he should address ways to assist in risk mitigation and factors that will further enhance an organization's business processes. A privacy officer should establish a risk-based approach to gain awareness of an organization's risk profile to maximize the policy potential.

- Expert in international privacy laws: Throughout various geographies, regulations go beyond local boundaries; hence, a privacy officer should know applicable global privacy laws. A skilled privacy officer helps an organization frame policies, identifies applicable privacy laws, and develops and implements adequate data protection schemas.

- Strong communication and verbal skills: A privacy officer works with various departments within an organization, such as the CISO and human resources department, to evaluate data protection procedures. While doing so, a privacy officer needs to communicate well within an organization to ensure the design of a data protection program meets the needs of the business.

- Knowledgeable about privacy implications and digital enhancements: Organizations are looking to integrate advancements within their overall operation, such as adopting technologies like cloud computing. A privacy officer must be aware of the confidentiality risks and implications associated with such technologies. In a shared services model, organizations usually share information with the service provider. Therefore, a privacy officer should be aware of the implications of information exchange to an organization using these services, including impact on data governance and the responsibility model.

- Increased knowledge of encryption technologies: As an organization grows, more and more data is shared through communication systems. A privacy officer needs to be well versed with data protection methodologies since the line between privacy and security is increasingly blurring. She or he should have increased knowledge of encryption technologies to efficiently protect bulk data shared and used in small grids and networks. In healthcare, for instance, patient data confidentiality and privacy are vital for a safe and secure care provision. Healthcare organizations require privacy officers to adopt data encryption technologies to ensure end-to-end data security and patient trust.

Skill sets and education required for a privacy officer

- A privacy officer should preferably have a bachelor's degree in the field of business; when applying for a senior-level position, a master's degree in privacy law or data protection may be preferred.

- She or he should aim to accomplish an International Association of Privacy Professionals (IAPP) certification.

- A privacy officer should have experience in guiding through data breaches and monitoring data compliance.
- She or he has expert knowledge of data protection policies and privacy laws.
- A privacy officer should have experience in running impact assessments specific to risk and privacy with a strong knowledge of data protection laws within respective jurisdictions of practice.
- She or he should have strong business acumen such as analytical, tactical and communication skills.

Annual compensation range (USD)

Compensation ranges vary globally, especially when comparing developed versus developing nations. To provide a statistical estimate based on relatively available data, the average compensation of a privacy officer within the United States is $85,000 USD. Most privacy officers earn between the range of $52,000 USD and $140,000 USD annually.[27]

CYBERSECURITY SALES AND MARKETING SPECIALIST

A cybersecurity sales and marketing specialist generates revenue for an organization by promoting security solutions or security services. This includes developing the marketing strategy and sales collateral; attending conferences to discuss established security systems and solutions; building a strong sales pipeline; following through and traversing respective sales cycles.

A cybersecurity sales and marketing professional also demonstrates to existing or potential clients about new and innovative security solutions that might help them counter emerging threats. An organization emphasizes hiring expert cybersecurity sales and marketing specialists to boost revenue and establish authentic customer relationships.

Key responsibilities of a cybersecurity sales and marketing specialist

- Understand the market and conduct market research: A cybersecurity sales and marketing specialist must bridge the market's security needs and an organization's security offerings; this includes leveraging a specialist team to proficiently communicate the recent market gap and demand for security systems. He or she also conducts in-depth market research to gain awareness about the latest market trends.

  A cybersecurity sales and marketing specialist researches market conditions in national, local or regional areas and accumulates data to estimate a security product or service's potential sales. He or she also strives to establish innovative marketing campaigns for an organization's newly developed security services or system.

To be more precise in market research, a cybersecurity sales and marketing specialist also gathers data about competitors, competitor price margin, net sales and distribution modes and marketing.

- Respond to client inquiries: A client is the most critical part of any sales-oriented business, and keeping a client satisfied translates to more sales revenue over time. A cybersecurity sales and marketing specialist must give particular emphasis to respond to a client's query timely. The individual should carefully outline the customer service strategy with specific responses and support seamless and effective client accessibility.

- Overlook various activities during promotional events: Promotion and marketing of a security system, solution or upgrade ensure that an organization or a client seeking to enhance its security tools is aware of a particular product launch. Product visibility and accessibility play a vital role in boosting sales. Hence, a cybersecurity sales and marketing specialist should help manage a promotional event to facilitate its success.

- Create and deliver promotional presentations: A cybersecurity sales and marketing specialist should aim to outline, supervise and execute marketing projects and promotions to drive awareness. She or he should efficiently prepare and give a promotional presentation that draws a potential client to invest in a newly launched security product.

- Utilize social media to its potential: Social media is a great platform to boost visibility and engage with potential clients. Cybersecurity sales and marketing specialists should aim to design innovative content that attracts client attention to boost sales. He or she should plan a social media calendar explaining the product and its benefits to attract clients into investing in an organization's security products with sales patterns top of mind. For example, a cybersecurity sales specialist should be acutely aware of peak and low sales cycles as well as business budget cycles to effectively time the delivery of social media promotions for when businesses are paying most attention.

- Outlining sales opportunities: In a highly competitive market, a cybersecurity sales and marketing specialist may also choose to associate with authorized channel partners to determine sales opportunities. These can include building a pipeline of opportunities through distributors and service providers. She or he also demonstrates the sales pitch and defines the expected sales and revenue.

Characteristics of a successful cybersecurity sales and marketing specialist

- Extensive market knowledge: A cybersecurity sales and marketing specialist should have in-depth knowledge of the current market, current events and market opportunities. This knowledge helps in providing logical suggestions to ensure security products are positioned appropriately. It is essential to pay attention to the latest market trends and customer requirements in a sales-oriented business.

- Good at communicating with a technical audience: A successful cybersecurity sales specialist is well-versed in communicating and persuading a technical cybersecurity audience. He or she can expect to be challenged and asked intricate questions about how the product or service delivers benefits to a technical audience who know their systems inside and out. She or he should adopt a comprehensive approach to demonstrating the security product and communicate well with these audiences.

- Goal-oriented skills to succeed in a dynamic environment: A cybersecurity sales specialist should have a focused vision and smart plan to execute the vision. Being goal-oriented and having a precise plan allows a cybersecurity specialist to track each achieved milestone. Anticipating the result helps in creating improved strategies to succeed in a dynamic environment.

- Highly observant: A successful cybersecurity sales specialist helps establish new and creative sales pitches for products. He or she should be detail-oriented and observe competitors' marketing strategies to frame a plan that helps boost sales.

- Ability to execute sales methodologies to generate increased revenue: A cybersecurity specialist should be well-aware of effective sales methodologies and quickly adapt and implement various methods when required. He or she should have a strategic and tactical approach to maximize an organization's revenue growth.

Skill sets and education required for a cybersecurity sales and marketing specialist

- She or he should have a college diploma or bachelor's degree in business administration or commerce with a concentration in marketing and sales or a relevant field.

- She or he should know the foundational concepts of cybersecurity and related regulatory practices.

- She or he should have strong interpersonal, communication and leadership skills.

Annual compensation range (USD)

Compensation ranges vary globally, especially when comparing developed versus developing nations. To provide a statistical estimate based on relatively available data, the average compensation of a cybersecurity sales and marketing specialist within the United States is $104,000 USD, with an average salary ranging between $21,500 USD and $181,000 USD.[28]

Cybersecurity Researcher

A cybersecurity researcher works within academia or organizations with specific cybersecurity research and investigations departments. A cybersecurity researcher has a solid educational background and works within his or her organization to discover and identify new and emerging solutions, theorems, and concepts within the field of cybersecurity. This research-driven role does not apply to all organizations.

Additional activities a cybersecurity researcher may be involved in include reverse engineering, network forensics, hypothesis development and building theorems and concepts for structured and unstructured learning to advance the field of cybersecurity. One such example is malware deconstruction to glean intelligence about a malware's behavioral characteristics—such as how it is structured and how it communicates—to help security analysts, malware specialists and incident responders detect instances of that malware in the future.

Key responsibilities of a cybersecurity researcher

- Summarize research results and contribute to system enhancements: A cybersecurity researcher actively interacts with the broader community by summarizing research results and publishing research papers in academic settings. He or she also shares knowledge within an organization by articulating recommendations through presentations and research papers to technical management, departments and government decision-makers. A cybersecurity researcher also aims to participate in sponsor meetings, community working groups, conferences and proposal writing.

- Enhance defensive capabilities: A cybersecurity researcher should actively collaborate with fellow experts to help develop defensive capabilities of critical internal systems. She or he should be able to analyze and evaluate current abilities and recognize exposed gaps. A cybersecurity researcher often works with security specialists to devise and establish new solutions and algorithms to enhance defensive capabilities.

- Frame and establish new systems: A cybersecurity researcher helps design and develop new applications, solutions and strategies by constantly observing evolving patterns and intrusion methods to effectively suggest appropriate solutions to limit future instances.

- Contribute to the enhancements to vulnerability detection tools: A cybersecurity researcher implements scientific rigor to add to improved comprehension of the cybersecurity domain. She or he helps in defining, describing and predicting behavior for determinations in cyber systems. A cybersecurity researcher should conceive and develop new methods, vulnerability detection tools, strategies or techniques to address crucial cybersecurity technical issues.

- Perform system test and anti-jamming: A cybersecurity researcher should conduct applied or basic research to identify urgent gaps in cybersecurity, including host-based security, network security, cloud security, mobile security, software development, integrated automated cybersecurity and IoT security. She or he should develop cybersecurity enhancements for malware detection, mitigation, isolation, prevention, visualization and analysis as well as perform dynamic testing, assessment, penetration, and anti-jamming.

Characteristics of a successful cybersecurity researcher

- Constant learner: A cybersecurity professional working in research should be passionate about learning new items and acquiring extensive knowledge. She or he is continually researching evolving security threats and ways to reverse their effects. A cybersecurity researcher contributes significantly to developing constructive and defensive tools. Security specialists in an organization rely upon research reports to determine the developed system's authenticity, strategies and tools.

- Curious and well-informed: A cybersecurity researcher should extract relevant information about potential vulnerabilities and threat actors. He or she is well-aware of methods to examine an organization's security operations from a holistic view, such as threat modeling, implementation, specification and vulnerability assessment. A cybersecurity researcher should also comprehend security problems related to networking, operating systems and virtualization software.

- Communication skills: A cybersecurity researcher is responsible for gathering logical and rational information to assist in developing systems, approaches and solutions. She or he presents final findings in front of executives and relevant business audiences. Hence, a cybersecurity researcher must have strong communication skills to present his or her conclusions and address inquiries effectively. A cybersecurity researcher should also have strong writing skills to create precise and well-structured reports.

- Enable the business: A cybersecurity researcher should aim to help an organization by staying ahead of the adversary by exploring and helping establish robust defense tools. She or he should look for grey areas in the security system and suggest practical solutions to deal with risky situations backed by thorough research.

Skill sets and education required for a cybersecurity researcher

- A cybersecurity researcher should have a bachelor's degree in computer science, cybersecurity or an equivalent diploma. Cybersecurity researchers working in academia will hold a master's or a Doctor of Philosophy (Ph.D.) in similar fields.

- She or he should aim to have extensive knowledge of security development methodologies, malware, network attack types and programming language.

- A cybersecurity researcher should have impactful business acumen and interpersonal skills.

Average compensation range (USD)

Compensation ranges vary globally, especially when comparing developed versus developing nations. To provide a statistical estimate based on relatively available data, the average compensation of a cybersecurity researcher within the United States ranges between $75,000 USD and $126,000 USD.[29]

TECHNICAL AND NON-TECHNICAL ROLES

Security Assessor/Assurance Officer

A security assessor or security assurance officer works within an organization with a predefined educational background and acquires certifications to meet the specific frameworks or regulatory requirements. The certifications help the assessor build credentials and recognition for his or her expertise. One of the more common security assessors is a certified Payment Card Industry Data Security Standard (PCI-DSS) assessor, known as a Qualified Security Assessor (QSA). A QSA reviews an organization's security program and ensures it conforms to the requirements laid down by the compliance mandate.

A security assessor is highly crucial for an organization. She or he identifies information security system gaps when mapping against industry standards or compliance regulations and suggests probable solutions to help the organization meet such standards and regulations. A security assessor may also contribute to the development of new strategies for the information security system.

Key responsibilities of a security assessor

- Benchmark systems against compliance requirements: A security assessor helps organizations bridge gaps between the current state and specific compliance mandate requirements. An in-depth analysis of the information system allows the assessor to understand the required remediation effort to achieve a compliant state. Additionally, in doing so, he or she helps draft sensible and concrete solutions to make the assessed system or process within an organization more grounded and robust against security vulnerabilities.

- Help build a competitive edge: Since the security assessor helps navigate the complicated road to compliance, he or she provides a sense of assurance that an organization is meeting the relevant global standards. In most cases, conformation to these compliance mandates is not optional. The security assessor helps the organization stay ahead of current and future compliance and security leading practices requirements.

- Matures security governance: While building a solid foundation for meeting compliance requirements, a security assessor also suggests the required reporting mechanisms, roles and responsibilities, and policies and procedures. An assessor ensures this structure remains applicable long term and reflects the changing dynamics of relevant standards and compliance.

- Risk view and collaboration: The technology and security field are ever-evolving. Hence, security experts should continually assess the system for deficiencies. A security assessor collaborates with an organization's risk unit to identify system insufficiencies and propose the information security system's requirements. This information flows both ways (i.e. the risk department might also suggest compensating controls that might fulfill a specific compliance requirement).

- Audit data through E-discovery: The fundamental objective of an organization is to protect confidential data against security abrasions. A security assessor defines strategies to protect specific data sets and, in some cases, audits that data set to ensure compliance. To protect data efficiently, a security assessor may identify critical data sets through the E-discovery process.

- Advisory services for security initiatives: A security assessor might also provide directions for security exercises in the system development life cycle (SDLC) and application advancement endeavors while also participating in authoritative undertakings as required.

Characteristics of a successful security assessor

- Understand the business model: A skilled security assessor should be acquainted with an organization's purpose and consequently its business model. Ample knowledge about the business' organizational structure helps the assessor cater to lagging areas with proficient amendments that help keep the organization meet its compliance and leading practice requirements.

- Expert in applicable security standard: A security assessor should be highly skilled in compliance with the specific security standard. For example, a PCI QSA should know the PCI world and translate those requirements to all stakeholders; this ensures the organization maintains its compliance responsibility, thereby instilling assurance to its consumers and regulators.

- Assessment scoping: For a security assessor to become proficient in his or her field, assessment scoping is vital. The assessor appropriately scopes the work at hand and provides a meaningful output that is translatable and usable for the business. Without proper scoping, an organization may not collect the right evidence or close the right remediation gaps.

- Suggest compensating controls: Compliance with industry standards can be a highly complicated task requiring timely remedial action, resulting in imposed financial and legal penalties if not conducted correctly. A skilled security assessor understands this and should be creative enough to suggest, where appropriate, existing or temporary controls can meet the compliance requirement for the time being until a more permanent solution is put in place. She or he also does this with appropriate rigor to not expose the organization to undue risks.

Skill sets and education required for a security assessor

- A security assessor should preferably have a university or college education in business, accounting or cybersecurity.

- She or he should aim to acquire security management certifications such as the CISSP and CISM. An auditing certification such as CISA is helpful.

- If a security assessor is focusing on a very particular standard or regulation, she or he should become certified and recognized in that domain. Examples may include ISO 27001 or PCI-DSS.

- A security assessor has detailed knowledge of the business units to understand security requirements and awareness of risk methodologies and fundamentals.

- She or he should have sufficient experience creating security architecture reports, documenting security plans and remediation strategies, working in close coordination with security architects and engineers.

- A security assessor can effectively communicate valuable and risk-based information to various audience levels across the organization, including senior management and executives.

- She or he should have project management experience to establish management plans such as resource allocation and budgeting.

Average compensation range (USD)

Compensation ranges vary globally, especially when comparing developed versus developing nations. To provide a statistical estimate based on relatively available data, the average compensation of a security assessor within the United States is $103,000 USD. Most security assessors earn between the range of $48,000 USD and $134,000 USD annually.[30]

CYBERSECURITY AUDITOR

A cybersecurity auditor is a well-informed IT professional who plays the influential compliance and controls assessment roles within an organization. She or he analyzes and scrutinizes the security controls and oversees an organization's management processes and security systems to ensure compliance with IT, OT and IoT standards. Overall, the cybersecurity auditor's tasks ensure an organization's processes meet individual requirements and follow leading security practices to help with risk and business alignment.

A cybersecurity auditor estimates the IT operational process efficiencies and compliance within an organization's security approaches correlated against regulations and compliance requirements. While the security auditor is not responsible for implementing or maintaining security controls, she or he can recommend possible enhancements based on security system reviews to ensure information systems security and integrity.

A cybersecurity auditor can evaluate deficiencies within the security system by running tests to evaluate the effectiveness of an organization's cybersecurity components. Many organizations rely on a cybersecurity auditor to provide well-documented security audits to determine and manage vulnerabilities and threats before potential risk or harm affects the information system.

Key responsibilities of a cybersecurity auditor

- Design and conduct security audits: A cybersecurity auditor should be proficient in formulating and executing security audits that fit an organization's business, policies and security approaches. She or he accesses and examines security practices and controls to ensure compliance with relevant security standards while working closely with managers, IT professionals and executives. A cybersecurity auditor also utilizes computer security expertise to run tests for risk and controls gap identification.

- Strengthen controls: A cybersecurity auditor should have ample knowledge regarding system vulnerabilities and identification. She or he should comprehend organizational processes, standards and policies to efficiently evaluate the deficiencies within a security system and suggest required enhancements. A thorough understanding of an organization's business objectives helps a cybersecurity auditor rationalize potential control enhancement to align with organizational goals.

- Evaluate the efficiency of security policies: A cybersecurity auditor should determine the compliance and efficacy of organizational operations synchronization with current security policies. She or he provides a multi-dimensional perspective about an organization's security practices. Organizations utilize a security auditor's expertise to investigate their security policies and ensure their current approach adheres to overall conformance to compliance and best practices.

- Interview personnel to determine security complications: A cybersecurity auditor conducts interviews with IT professionals, managers and executives to understand existing control efficiencies and establish strategies for remediating gaps. These strategies are proficient in enhancing security compliance, handling potential security vulnerabilities and managing risk.

- Execute proper documentation of the audit process: The most fundamental responsibility of a cybersecurity auditor is to properly document the overall audit process on various computer applications and computing environments. Auditing organizational security and networking systems are required for security efficiency. Then the evaluated results must be added and documented to frame the final audit report. In most cases, the audit report is the single source of truth regarding existing controls and how those controls are commensurate to applicable compliance requirements.

Characteristics of a successful cybersecurity auditor

- Analytical and critical thinking skills: A cybersecurity auditor should have polished critical thinking and analytical skills to establish tests grounded on organizational policies effectively. She or he should implement industry standards using strong analytical and critical thinking abilities and develop comprehensive tests that efficiently fit an organization's security policies.

- Strong writing and verbal abilities: A cybersecurity auditor conducts interviews with executives and employees to extract information and build strategies. This information helps create detailed audit reports. She or he relies upon the findings of interviews and workshops to suggest enhancements, improvements, and gap remediation. A cybersecurity auditor develops concise documents addressing potential weaknesses and gaps in the security system. Impactful written and verbal communication skills are required to carry out these tasks efficiently.

- Passion for learning: A cybersecurity auditor should be passionate about gaining additional knowledge regarding information systems and computer applications of diverse complexity. She or he should be able to comprehend industry-specific information and security regulations effectively. A security auditor working in insurance or healthcare organizations should ensure she or he complies with the concerned accountability act and portability. A security auditor should strive to gain knowledge of relevant programming languages and penetration tests as it further enhances his or her ability to positively contribute to an organization's success.

- Strategic thinker: A cybersecurity auditor must be a strategic thinker and highly adaptive. Ever-evolving technology, legal matters, internal controls and changing socio-economic conditions influence an organization's competitiveness. An experienced cybersecurity auditor identifies the impact of these external forces and keeps them in mind while testing the internal system for conformity and risk. The ability to think strategically helps a cybersecurity auditor analyze all aspects of a security system to draw effective outcomes.

Skill sets and education required for a cybersecurity auditor

- A cybersecurity auditor should preferably have a bachelor's degree in information technology, computer science or another related field. The field of accounting and finance also has specific transferrable skills into auditing.

- She or he must aim to obtain certifications such as a Certified Information Systems Auditor (CISA). Additional cybersecurity leadership certifications applicable to this role include the CISSP and Certified Risk and Information Systems Controls (CRISC).

- A cybersecurity auditor should have detailed knowledge of standard IT processes, well-known tools used within an organization.

Average compensation range (USD)

Compensation ranges vary globally, especially when comparing developed versus developing nations. To provide a statistical estimate based on relatively available data, the average compensation of a cybersecurity auditor within the United States ranges between $80,500 USD and $171,000 USD.[31]

Security Consultant

A security consultant is an advisor who manages and provides a vast array of security-related services to clients. He or she mainly works with an organization's risk, compliance, information technology and security department to assist these teams on their security initiatives. A security consultant has strong business acumen and extensive technical knowledge, empowering him or her to provide subject matter expertise to disparate security and risk scenarios.

An organization that is continuously facing difficulties concerning its security system relies on security consultants to provide subject matter expertise. A security consultant offers his or her expertise and is responsible for safeguarding sensitive information, critical data sets, financial assets and training the internal security teams to ensure the organization's security systems, processes and resources are robust.

Key responsibilities of a security consultant

- Security advisory services: A security consultant is a subject matter advisor (SMA) who provides expertise to help manage IT risks within the risk threshold. To viably manage this, security consultants should be able to develop dynamic strategies. A security consultant should analyze the security program and assess whether current plans and initiatives within an organization adequately address risks.

- Establishing security plans for clients: A security consultant should consider the security needs of a client. He or she creates a sustainable security plan through a top-to-bottom analysis of the client's security framework by reviewing and understanding the threats to the organization.

- Sales pursuits: Depending upon business and seniority, a security consultant might require maintaining an updated pursuit funnel. A security consultant contributes to sales of security services and invests time understanding the competitor's landscape, essential buyers, procurement channels and liaise with sales pursuit leads.

- Ideate and innovate: Threat actors rarely use the same approach to break into an organization's system. Consequently, a security consultant should continuously keep up with the latest protection methods and the adversary's latest modus operandi.

Characteristics of a successful security consultant

- Negotiation skills: A security consultant frequently needs to guide his or her potential clients about their services. When trying to land a contract with a client, negotiation skills play a vital role. A security consultant should have adequate negotiation skills to seal the deal on preferred terms.

- Communication abilities: A security consultant interacts with many people, including employees and clients. When conversing with a client about security services, a security consultant should have sharp communication skills to work with all levels of a client's organization as a trusted security advisor and to form long-term relationships. A security consultant, likewise, educates team members about security arrangements and recently identified issues.

- Leadership skills: A security consultant is also responsible for educating and managing a team of employees to establish security strategies while efficiently building and managing a security program. The consultant needs to have influential leadership and coaching abilities to motivate team members and align them with the overall organizational vision. A successful security consultant knows how to use practical leadership skills to ensure team members are on the same wavelength and collectively move forward to achieve an organization's objectives.

- Continuous learning: A security consultant should be quick in providing an innovative solution to cater to the newest security issues. He or she should be eager to learn new techniques and maintain ongoing curiosity to gather information about new strategies that can assist with building improved security plans.

- Perceptive and curious: A security consultant maintains a high level of curiosity to excel in a cybersecurity-related field. He or she should be able to anticipate deficiencies before an occurrence in a security system. This quality encourages a security consultant to handle such circumstances effortlessly, and she or he can glance through the threat actors' lens to predict their next move.

- Critical thinking: A security consultant effectively structures, creates and implements contingency protocol with a security program. He or she should think critically to guide clients with risk definitions, risk alignment and program sustainability. The capacity to think critically helps a security consultant effectively handle alarming situations and establish rational solutions to keep vulnerabilities under control.

- Patience: In the world of cybersecurity, determination and tolerance is the key. A security consultant must manage numerous concerns simultaneously, and this can be exhausting. A security consultant should be patient in case of emergencies to adequately address the issues and lead and/or support his or her team.

Skill sets and education required for a security consultant

- A security consultant should have a bachelor's degree, preferably in IT, computer science or cybersecurity.

- He or she should strive to achieve certifications such as CEH, CISA, CISSP or CISM to validate his or her abilities to establish and reinforce security programs.

- A security consultant should have solid interpersonal and communication skills.

- She or he must have strong business acumen, including translating technical jargon into business language and vice versa.

Average compensation range (USD)

Compensation ranges vary globally, especially when comparing developed versus developing nations. To provide a statistical estimate based on relatively available data, the average compensation of a security consultant within the United States is $86,000 USD. Most security consultants earn between the range of $61,000 USD and $144,000 USD annually.[32]

CYBERSECURITY INSTRUCTOR AND TRAINER

A cybersecurity instructor is an educator and a coach at the core. She or he is a professional who develops the curriculum and conducts classes and sessions to guide cybersecurity learners. A cybersecurity instructor is a well-researched and well-informed professional who strives to bring real-world scenarios into the classroom.

A cybersecurity instructor conducts detailed training to guide students about the latest developments in the field of cybersecurity. She or he mainly works in post-secondary institutions, private training institutes or part of an organization's training department. A cybersecurity trainer must learn the latest tools, technologies and methodologies to combat security risk and is ready to brainstorm and ideate in a classroom setting.

A cybersecurity instructor coaches aspiring cyber professionals and guides them to develop rational approaches and strategies for limiting risks. A cybersecurity instructor working in an educational institute develops an applicable curriculum, creates learning materials and assesses student skills and development pace. She or he should also look out for innovative and intuitive ways to impart knowledge by augmenting curriculum-based learning with real-life scenarios. A cybersecurity instructor should establish a learning atmosphere to foster intellectual curiosity within students, which induces strategic thinking to solve complicated issues.

Key responsibilities of a cybersecurity instructor and trainer

- Provide training sessions: The utmost priority for a cybersecurity instructor and trainer is to impart the appropriate knowledge to aspiring professionals using delivery methods that provide the most intuitive experience. During training, a cybersecurity instructor and trainer provides a broader industry perspective while the delivery stays close to the curriculum. When working at an educational institute, the ideal method is to augment hands-on exercises where applicable (using lab space) to conduct practical classes, give knowledgeable lectures and provide theoretical guidance through notes and tutorials.

- Facilitate coursework as per an institution's core values: A cybersecurity instructor and trainer should be mindful and establish comprehensive ways to integrate practical learning in synchronization with an institution's core values. She or he should manifest a safe and professional learning environment by setting the tone of the classroom and modeling good workplace behavior and skills. A cybersecurity instructor and trainer should establish a positive learning environment by enforcing classroom management and observing student code of conduct.

- Keep up and impart knowledge about the latest advancements: A cybersecurity instructor and trainer should strive to absorb the emerging information to present updated technical awareness to students or the cybersecurity team. She or he should aim to observe the latest threat patterns in an organization's security system to mitigate risk effectively. On the other hand, when conducting educational classes, the cybersecurity instructor must deliver the latest technical knowledge to enhance learning.

- Establish an atmosphere conducive to learning: A cybersecurity instructor and trainer should foster a positive environment to help students reach their educational, career and personal goals. The learning environment should encourage diverse ideas by modeling respect and stimulating confidence to improve each student's ability to learn. A cybersecurity trainer should employ various summative and formative evaluations to ensure adequate learning.

- Statistics and predictive analysis: A cybersecurity instructor and trainer should guide students or cybersecurity teams about predictive analytics and statistics. Predictive analytics helps identify and halt various criminal behaviors before inflicting any system damage. Predictive analytics can assist in examining user actions and responses. An organization can efficiently detect suspicious activities such as corporate spying, credit frauds and cyberattacks by incorporating predictive analysis and analyzing the statistics. Knowing this allows a cybersecurity instructor to build hypotheses and examples from the real world.

- Mentorship: A cybersecurity instructor and trainer is often also recognized as a mentor for learners. There are often aspiring professionals who need specific guidance on education, career paths and networking, and the instructor can provide that information. This book covers the importance of mentorship in Chapter 7.

Characteristics of a successful cybersecurity instructor and trainer

- Have a passion for teaching and advancing the field: A cybersecurity instructor and trainer is usually the first interaction an aspiring cyber professional has with cybersecurity. A cybersecurity instructor and trainer should have a strong passion for the field exhibited in her or his personality and pass through a positive influence on learners.

- Knowledge of security tools: To maintain a sound security system, a cybersecurity instructor or trainer must be aware of the industry's latest security tools and the applicability of those tools. An organization's cybersecurity instructor and trainer must bring security tools knowledge to conduct training sessions. Learning security tools helps provide learners an overview of what is working from a controls perspective regarding technology. A cybersecurity instructor and trainer who gain influential standing in the field must strive to attain the latest security tool knowledge to help students or cybersecurity teams establish comprehensive system understanding.

- Knowledge of collaboration tools and knowledge management: A cybersecurity instructor and trainer must also impart knowledge about collaborating and sharing research information securely. In that light, she or he should have extensive knowledge of collaboration tools to provide for sufficient communication guidance in sharing current organizational content internally. Many organizations have knowledge exchange platforms and repositories where employees share procedures and guidelines. A trainer should advocate this platform and ensure that it is up-to-date and easy to use.

- Knowledge of emerging technologies: A cybersecurity instructor and trainer should seek information about emerging technologies, such as cloud-based security, endpoint security and next-generation security infra-structures. Knowing the latest technology helps a cybersecurity instructor and trainer provide sufficient guidance to students or cyber teams to efficiently in-tegrate technology that is robust in identifying and protecting systems against risk.

Skill sets and education required by a cybersecurity instructor or trainer

- A cybersecurity instructor and trainer should have a bachelor's degree in computer engineering, computer science or another relevant field. A senior profes-sional should aim to obtain a master's degree in similar areas focused on research.

- She or he must target acquiring relevant professional cybersecurity certifications, such as the CISSP, CISM and others.

- A cybersecurity instructor should have detailed knowledge of the latest advancements, data analytics and collaboration tools. Knowledge of these is ob-tained through constant learning and becoming a member of several knowledge-sharing forums and professional groups, such as the (ISC)^2 chapters.

Average compensation range (USD)

Compensation ranges vary globally, especially when comparing developed versus developing nations. To provide a statistical estimate based on relatively available data, the average compensation of a cybersecurity instructor within the United States is $96,000 USD, with average salary ranging between $64,500 USD and $114,000 USD.[33]

CHAPTER 4 KEY TAKEAWAYS

This chapter's primary objective is to provide a detailed view of the various cybersecurity roles, an overview of critical responsibilities, skill set requirements and an indicative compensation range. There are certain variations or combinations of these roles within different organizations and various parts of the world. It is essential to understand that larger organizations may have 1:1 mapping to the described roles. In contrast, smaller organizations may combine a few roles per their budget and business needs.

Takeaways:

- Cybersecurity is not just for technical professionals; it is a vast field requiring diverse skills and expertise. There are managerial, technical, non-technical and blended roles for consideration.

- There are various references available to augment the information shared in this book about cybersecurity roles. One commonly known cybersecurity workforce framework is the NIST National Initiative for Cybersecurity Education (NICE) Cybersecurity Workforce Framework, or the NICE Framework for short. It includes seven categories of common cybersecurity functions, 33 specialty areas and 52 work roles. This framework has become the cyber work role lexicon and structure for companies, job seekers, recruitment agencies and certification organizations. In this chapter, we explored some and not all cybersecurity roles identified within the NICE Framework. For more information, visit https://niccs.us-cert.gov/workforce-development/cyber-security-workforce-framework.

- CyberSeek partnered with NIST, CompTIA and Burning Glass Technologies to create an interactive cybersecurity career tool. It includes a heat map of cybersecurity supply and demand in the U.S. and an interactive career pathway showing the common roles within cybersecurity and opportunities to transition between them. For more, go to www.cyberseek.org.

CHAPTER 5: WHICH CERTIFICATES SHOULD I PURSUE?

Becoming certified means a practitioner has demonstrated the level of competency, knowledge base, experience and ethics set by a professional organization's global standards. As such, recruiters and relevant stakeholders are assured a certified practitioner has successfully met these standard requirements when recruiting, hiring and promoting talent. Now more than ever in the field of cybersecurity, it is vital for a practitioner to keep up to date with the ever-changing technology and the demands of businesses and regulators to stay competitive.

IT certifications can primarily be based on two bodies of knowledge: vendor-specific certifications and vendor-neutral certifications. The former is a subset where a specific technology vendor forms the guidelines of what it will take to be certified specific to their products, including learning the Common Body of Knowledge (CBK), passing the exam and maintaining the certification. For example, the "Microsoft Certified: Azure Security Engineer Associate" is a cloud security certification created by Microsoft and is specific to the Azure knowledge base. The latter is technology agnostic and is often managed by a professional organization built on foundational knowledge. For example, the Information System Security Certification Consortium, commonly known as (ISC)[2], governs the Certified Information Systems Security Professional (CISSP) certification. The CISSP is based on foundational security knowledge, is widely accepted and is agnostic to technology, compliance regulation or industry.

According to infosec-careers[34], the following are the top 10 certifications that are currently in high demand:

1. CISSP – Certified Information Security Systems Professional, (ISC)^2
2. CISM – Certified Information Security Manager, ISACA
3. CEH – Certified Ethical Hacker, EC-Council
4. CompTIA Security+
5. OSCP – Offensive Security Certified Professional, Offensive Security
6. CCSP – Certified Cloud Security Professional, (ISC)^2
7. ISO/IEC 27001 Lead Implementer
8. APT – Advanced Penetration Testing, EC-Council
9. OSCE – Offensive Security Certified Expert, Offensive Security
10. GSEC – GIAC Security Essentials, GIAC

There is a wide range of cybersecurity certifications. Therefore, it is essential to take the time to clearly define her or his career aspirations and then determine which certifications are the right ones to help achieve that career. There is a common misconception cybersecurity professionals must have several certifications. The number of certifications alone does not necessarily equate to success; several other initiatives are also important, including finding a mentor and networking. Ultimately, overall success in a cybersecurity career focuses on gaining a reputation as a trusted security practitioner who brings business value to an organization.

That said, to become a trusted cybersecurity practitioner, obtaining the right certifications is one of the most critical decisions to make. Since obtaining any certification requires a substantial investment of time and money, the following criteria should be considered:

- Career level (junior, experienced, advanced)

- Security specialization

- Technical vs. non-technical background

- Understanding of the certification, level of difficulty of the exam, work experience required to meet the certification and ability to maintain the certification long term

- Available time to commit to studying

- Commitment to sustain the cost of certification

The certifications covered in this chapter are non-vendor specific from (ISC)^2, CompTIA, ISACA, EC-Council, Offensive Security and GIAC. All available certifications offered by these organizations are briefly explored in this chapter. While there are numerous vendor-specific certifications, these are intentionally not covered within this book because those certifications often change with vendor product strategies, certification and recognition approaches, and mergers and acquisitions in the marketplace.

(ISC)²

The International Information System Security Certification Consortium (ISC)² is globally known for its cybersecurity certification excellence. It is a not-for-profit organization that works on the vision of rendering valuable certification programs and standardization to shape information security practitioners. (ISC)² is an international membership organization providing a consolidated platform to cybersecurity professionals for on-demand certifications in varying domains.

(ISC)² has successfully catered to almost 150,000 cybersecurity professionals empowering them as international security officials. The consortium trains professionals using exemplary techniques and teaches new skills to elevate expertise. (ISC)² provides nine certifications, and each certification provides a specific bespoke knowledge base for cybersecurity professionals to grow professionally.

The nine (ISC)² certifications briefly covered in this chapter include:

1. CISSP
2. CISSP-ISSAP
3. CISSP-ISSEP
4. CISSP-ISSMP
5. SSCP
6. CCSP
7. CAP
8. CSSLP
9. HCISSP

Further details about (ISC)² are available on their official website at https://www.isc2.org.

CISSP

The Certified Information Systems Security Profession (CISSP) is an expert certification and is also the most sought-after credential among security professionals, recruiters and businesses globally. Many cybersecurity job postings, for instance, include the CISSP certification as an asset.

The CISSP is built upon a foundational Common Body of Knowledge (CBK) with eight information security domains augmented with leading world scenarios. The CISSP CBK effectively frames an outline concerning information security principles and concepts, enabling the professional to understand, debate and articulate potential solutions for real-world security risks.

To obtain a CISSP, in addition to passing the exam, the candidate must demonstrate at least five years of paid work experience in two or more of the eight domains of the CISSP CBK. A candidate can obtain a one-year credit out of the required five if she or he has a four-year bachelor's degree or an additional credential from the (ISC)^2 approved list.

Candidates who cannot demonstrate the minimum work experience can still write the CISSP exam, and once passed, she or he receives the Associate of (ISC)^2 designation. (ISC)^2 allows up to six years to earn the five years required to convert the Associate to the full-blown CISSP designation. Through (ISC)^2, the CISSP certification offers a professional with a wide array of educational tools, peer-to-peer opportunities for networking and exclusive resources.

Additionally, a candidate must comply with the CISSP code of ethics. The CISSP certification is only valid for three consecutive years. The candidate must choose to either appear in the exam again to renew the certification or decide to earn a well-prescribed set of continuing professional education (CPE) to qualify for recertification.

For more information, please refer to https://www.isc2.org/Certifications/CISSP.

CISSP-ISSAP

The Information Systems Security Architecture Professional (CISSP-ISSAP) is an elite, advanced-level certification issued by the (ISC)^2 focused on information security architecture. The CISSP-ISSAP CBK uplifts a professional's knowledge about architecture governance, architecture modeling, IAM architecture, AppSec architecture and security operations architecture.

In addition to passing the exam, to qualify for the CISSP-ISSAP, a candidate must be a CISSP in good standing and have two years cumulative, paid work experience in one or more of the six domains of the CISSP-ISSAP Common Body of Knowledge (CBK). A CISSP-ISSAP candidate must recertify every three years or, alternately, earn 2o Continuing professional education (CPE) credits each year focused on security architecture practice.

For details, go to https://www.isc2.org/Certifications/CISSP-Concentrations.

CISSP-ISSEP

(ISC)^2 introduced the Information System Security Engineering Professional (CISSP-ISSEP) certification in 2014 as an initiative to enhance a cybersecurity professional's abilities and career advancement specific to security engineering. It is among the most valuable specialized certifications provided by the (ISC)^2 organization as it promotes the comprehensive engineering skills of information security professionals.

CISSP-ISSEP certification mainly focuses on demonstrating knowledge regarding system security engineering, risk management framework, technical management, standards, policies and approaches. The CISSP-ISSEP certification uplifts a security engineer's skills by testing his or her abilities to design a logical security system.

In addition to passing the exam, to qualify for the CISSP-ISSEP, a candidate must be a CISSP in good standing and have two years cumulative, paid work experience in one or more of the five domains of the CISSP-ISSEP CBK. A CISSP-ISSEP candidate must recertify every three years or, alternately, earn 2o Continuing professional education (CPE) credits each year focused on security engineering.

The candidate must be able to frame logical security systems with proper approaches as defined by an organization. The applying professional should be able to implement security assessments to check for vulnerabilities adequately. The pattern for the examination is predefined to provide a comprehensive overview.

For more information, go to https://www.isc2.org/Training/Courses/issep-training-course.

CISSP-ISSMP

The Information Systems Security Management Professional (CISSP-ISSMP) focuses on the management aspects of information security. This certification integrates a wide array of knowledge to enhance an information security professional's management and leadership abilities. The CISSP-ISSMP certification incorporates security project planning and management, resiliency, designing continuity and response.

The CISSP-ISSMP certification covers five domains that involve security management and leadership, security compliance management, security life cycle management, contingency management and the law of incident and ethics management. CISSP-ISSMP is among the top ten most influential cybersecurity certifications substantiating that the candidate has experience in the various management-focused knowledge areas.

In addition to passing the exam, to qualify for the CISSP-ISSMP, a candidate must be a CISSP in good standing and have two years cumulative, paid work experience in one or more of the six domains of the CISSP-ISSMP CBK. A CISSP-ISSMP candidate needs to recertify every three years or, alternately, earn 2o Continuing professional education (CPE) credits each year.

Further details are available at https://www.globalknowledge.com/de-de/certifications/certification-training/isc/cissp-issmp.

SSCP

The System Security Certified Practitioner (SSCP) certification targets the idea of multi-variate approaches in managing underlying security administration and operations topics. SSCP advances the technical skills and fundamental knowledge for a security practitioner to monitor, administer and implement a secure IT infrastructure by utilizing the best security practices.

(ISC)^2 integrates leading policies and practices, well-accepted by cybersecurity professionals, to establish a comprehensive understanding of complex security issues. The SSCP certification is beneficial for information security professionals, network security professionals, security analysts, security administrators and database administrators.

To achieve the SSCP certification, a candidate must obtain at least one year of paid experience within at least one of the seven domains of the SSCP CBK. A candidate with a relevant degree from an accredited college or university is qualified for an experience waiver of one year. The candidate must abide by the code of ethics released by (ISC)^2 to avoid uncertain circumstances.

For more information, go to https://www.isc2.org/Certifications/SSCP.

CCSP

In the world of constant change, a cybersecurity professional has to face unique challenges frequently. A professional must strive to obtain appropriate, relevant and future-proof skills continuously. The Certified Cloud Security Professional (CCSP) is a leading, globally accepted certification catering to cloud security and its intricacies. The CCSP certification empowers a cybersecurity professional by demonstrating that she or he can help organizations tackle the cybersecurity complexities inherent in the ubiquitous migration-to-cloud efforts. No matter if an individual works for a cloud service provider or a cloud consumer organization, the certification demonstrates that an individual can speak the specific technical and governance language and can champion cloud-facing infrastructure security.

The CCSP certification is known for its global excellence and its ability to validate a cybersecurity professional's cloud security practices and approaches. The main advantages of the CCSP certification include a holistic approach in establishing instant credibility; demonstrating cloud security techniques using the most practical technology solutions; knowledge regarding versatile cloud security platforms; and a platform to advance a career in cybersecurity.

In addition to passing the exam, to qualify for the CCSP, a candidate must have a minimum of five years of paid experience in information technology, out of which three years must be in information security. Additionally, a candidate should also have one year in one or more of the six domains of the CCSP CBK. For candidates with a CISSP in good standing, the entire CCSP experience requirement is waived. If a candidate has a Certificate of Cloud Security Knowledge (CCSK) in good standing, one year is waived.

For more information, go to https://www.isc2.org/Certifications/CCSP.

CAP

A security professional requires ongoing knowledge to counter security threats and uncertain situations. The CAP – Security Assessment and Authorization certification is ideal for information security and assurance professionals to demonstrate their understanding and expertise within the Risk Management Framework (RMF). An information security professional with the CAP certification can show she or he comprehends an organization's security needs at the RMF level. The comprehensive understanding of an organization's mission and associated risks help a CAP-certified security practitioner integrate impactful security policies to catalyze control implementation assurance.

An information security analyst, information security specialist, information security manager or an information security auditor can apply for CAP certification to demonstrate her or his skills and gain support from the (ISC)^2 community.

In addition to passing the exam, to qualify for CAP, a candidate must have a minimum of two years of paid experience in one or more of the seven domains of the CAP CBK. Like any other (ISC)^2 certification, the individual has an option to successfully pass the CAP examination and obtain the Associate of (ISC)² if they do not have the required experience. And Associate of (ISC)² has the option to get the necessary two-year experience within three years of becoming certified to convert the associate designation to the CAP certification.

For more information, go to https://www.isc2.org/Certifications/CAP.

CSSLP

The Certified Secure Software Lifecycle Professional (CSSLP) certification is one of the most popular certifications related to software development. It demonstrates a cybersecurity professional's in-depth understanding of integrating security requirements into each phase of a Software Development Lifecycle (SDLC).

The CSSLP certification provides a cybersecurity professional a holistic overview of best practices for software establishment, application, testing and deployment. The CSSLP holds fundamental importance for project managers, software program managers, security managers, IT managers and quality assurance managers. The later a security flaw is discovered in the SDLC, the more expensive, time-consuming and complicated it becomes to address appropriately. Hence, software development programs must drive security as an integrated process weaved throughout the software development tasks.

In addition to passing the exam, to qualify for the CSSLP, a candidate must have a minimum of four years of professional SDLC work experience in one or more of the eight domains of the CSSLP CBK. Individuals with a four-year degree can waive one year from the required four years. Like any other (ISC)² certification, the individual has the option to successfully pass the CSSLP examination and obtain the Associate of (ISC)² if she or he does not have the required experience. The associate then can get the necessary four-year experience within five years to convert the associate designation to the CSSLP certification.

For more information, go to https://www.isc2.org/Certifications/CSSLP.

HCISPP

The Healthcare Information Security and Privacy Practitioner (HCISPP) certification is a great way to substantiate and assert experience in protecting patient health information (PHI) within the highly regulated healthcare sector. The HCISPP is an appropriate amalgamation of privacy techniques and practices with cybersecurity skills. The HCISPP provides the healthcare professional with an extensive overview of policies and best practices for establishing strong privacy protection models of PHI.

The HCISPP CBK provides comprehensive knowledge to enhance a professional's ability to manage, access and implement privacy and security controls. A healthcare organization with certified employees can frame logical approaches to protect PHI as it passes through the digital care continuum. A compliance manager, risk analyst, compliance auditor, health information manager or medical records supervisor should find the HCISPP certification highly applicable to advance his or her career.

A healthcare security professional applying for the HCISPP certification examination must satisfy specific standards. The fundamental criterion is a candidate must have at least two years of paid cumulative experience within the knowledge areas of the HCISPP CBK. Additionally, one of the two years above must be within the healthcare industry. Like any other (ISC)^2 certification, the individual has the option to successfully pass the HCISPP examination and obtain the Associate of (ISC)^2 if she or he does not have the required experience. The associate then can get the necessary two-year experience within three years to convert the associate designation to the HCISSP certification.

For more information, go to https://www.isc2.org/Certifications/HCISPP.

CompTIA

The Computing Technology Industry Association (CompTIA) is a non-profit IT professional association with leading IT certification programs. CompTIA serves in more than 120 countries providing vendor-neutral certifications for many years. The vendor-neutral certifications offered by CompTIA caters to more than 50 types of industries. CompTIA actively tracks and investigates the alteration within the IT industry over time and evaluates its certifications accordingly. CompTIA has issued over 2.2 million certifications since its establishment.

The four CompTIA security certifications briefly covered in this chapter include:

- Security+
- CySA+
- PenTest+
- CASP

Further details about CompTIA are available at https://www.comptia.org.

Security+

CompTIA Security+ is a globally recognized baseline security certification. It validates the foundational skills necessary for an individual to pursue a rewarding IT security career. CompTIA Security+ helps prove that a candidate has world-class, hands-on and foundational skills in security risk assessment, security incident monitoring, alerting and response. The Security+ certification may be an ideal certification for a candidate entering the cybersecurity field.

The Security+ certificate provides employers and peers a sense of confidence that a security professional has foundational knowledge about IT audits, system administration, security administration, network administration and penetration testing. Security+ incorporates hands-on troubleshooting as well as best practices to ensure a security professional gains practical problem-solving skills. Additionally, the Security+ certification is ISO compliant and approved by the U.S. Department of Defense (DoD).

To acquire the Security+ certification, a cybersecurity professional must comply with all the requirements outlined by CompTIA. Although there are no mandatory educational or professional experience requirements, CompTIA recommends two years of IT administration experience with a security focus. The minimum passing threshold for CompTIA Security+ is 750 marks (on a scale of 100-900). The certification is valid for three years and can be extended in blocks of three years by fulfilling the CompTIA continuing education requirements.

For more information, go to https://www.comptia.org/certifications/security.

CySA+

The CompTIA Cybersecurity Analyst (CySA+) is an intermediate-level certification with a strong focus on behavioral analytics of devices and networks to prevent, detect and thwart security threats through continual security monitoring and alerting. The CySA+ certification emphasizes current protection capabilities such as threat hunting, orchestration and mapping intricacies of network traffic patterns and user behavior patterns.

This certification supports a security professional's core abilities by integrating knowledge areas such as efficient data interpretation; determining and identifying vulnerabilities; utilizing threat detection measures; and extracting quick solutions. CySA+ is an applicable and differentiating certification for cybersecurity professionals such as incident handlers, threat intelligence analysts, compliance analysts, threat hunters and application security analysts.

To apply for the CySA+ certification, a candidate must fulfill the requirements outlined by CompTIA. Although there are no mandatory experience and education requirements, CompTIA recommends that a candidate have direct or indirect experience in the information security field. There are no other prerequisites for the CySA+ certification other than those mentioned above. The CompTIA general examination policy defines the passing threshold and the time limit. The certification is valid for three years and can be extended in blocks of three years by fulfilling the CompTIA continuing education requirements.

For more information, go to https://www.comptia.org/certifications/cybersecurity-analyst.

PenTest+

Penetration testing is crucial in protecting an organization's security system and determining potentially exploitable situations. The CompTIA PenTest+ is a unique certification requiring a candidate to demonstrate hands-on ability and knowledge about system controls and testing potential exploits in a controlled environment. Establishing and operating a threat and vulnerability management (TVM) program is one of the most operationally intensive and critical security operations functions.

The PenTest+ certification establishes the substantial ground for a cybersecurity professional to acquire and implement best practices and improve offensive security practices.

PenTest+ is an exemplary certification as it incorporates viable techniques to integrate and demonstrate knowledge and the ability to perform hands-on tests on security controls. This certification assures a security professional can perform hands-on controls testing to ensure the security configurations for mobile, desktop servers and cloud services can withstand today's attacks.

A candidate must understand the compliance requirement and required skills to appropriately scope and plan a controls assessment. To be successful in the exam, a candidate must understand the most up-to-date knowledge about tools and techniques regarding penetration testing, vulnerability assessment, data analysis, proper reporting and results communication. CompTIA recommends a candidate have three-to-four years of hands-on security controls testing or related experience, although there are no formal mandatory experience requirements. The certification is valid for three years and can be extended in blocks of three years by fulfilling the CompTIA continuing education requirements.

For more information, go to https://www.comptia.org/certifications/pentest.

CASP+

The CompTIA Advanced Security Practitioner (CASP+) certification caters to experienced security professionals who chose to remain technically inclined as his or her career progresses, as opposed to evolving into leadership roles. The CASP+ exam focuses on testing relevant advanced technical security skills instead of strategic managerial topics. The CASP+ certification validates a professional's technical prowess in risk management, security architecture, design, security operations, resiliency and technical controls implementation.

The CASP certification narrows its focus to the technical security domains of architecture and design concepts, risk analysis through trend data interpretation and cyber defense capabilities aligning with business goals. With the CASP certification, a security professional advances her or his understanding of security controls concepts, software vulnerability evaluation, and the principle of cryptographic technique implementation, such as mobile device encryption, blockchain and cryptocurrency.

To apply for CASP certification, a candidate must fulfill the standard requirements issued by the organization. The CompTIA recommends a candidate have a minimum of 10 years of experience in IT administration, five of which should be in technical hands-on security roles. However, there are no formal mandatory experience requirements. The certification is valid for three years and can be extended in blocks of three years by fulfilling the CompTIA continuing education requirements.

For more information, go to https://www.comptia.org/certifications/comptia-advanced-security-practitioner.

ISACA

The Information Systems Audit and Control Association (ISACA) is a leading global professional organization, incorporated in 1967, working actively to advance the best talent, expertise, and learning in technology. ISACA operates in more than 180 countries and serves 145,000 professionals, spanning strategic roles in IT governance, assurance, risk and cybersecurity. Through the years, ISACA has established highly sought-after IT risk, governance and information security certifications and has built widely adopted IT frameworks, like the Control Objectives for Information and Related Technologies (COBIT).

Through its global network of highly engaged professionals, ISACA helps catalyze highly innovative collaboration to solve today's information protection challenges. ISACA also operates local chapters bringing together IT risk, audit, governance and security professionals to engage within geo-specific technology protection challenges. It also provides access to valuable knowledge resources through published content or digital delivery for career growth. The continuous evolution within the cybersecurity field requires a professional to be nimble in acquiring relevant skill sets and collaborating with like-minded professionals. ISACA strives to keep its certifications up-to-date and closely mapped with knowledge areas at the forefront of addressing today's threat landscape.

The nine ISACA certifications briefly covered in this chapter are listed below:

1. CISA
2. CISM
3. CRISC

4. CGEIT
5. CSX-P
6. CDPSE

Further details about the certification are available at https://www.isaca.org.

CISA

The Certified Information System Auditor (CISA) is one of the most popular IT certifications that could be a significant differentiator in a cybersecurity professional's career trajectory. More than 150,000 professionals have obtained the CISA certification, which is a key indicator of the certification's popularity.

This certification is highly sought-after by information system auditors, governance professionals, controls assessors and compliance leads to name a few. The CISA certification validates a professional's ability to assess the effectiveness of security controls and governance processes of information systems. The CISA CBK seeks to integrate contemporary field-tested techniques and knowledge, which comes in very handy while preparing for the certification. The CISA CBK provides a professional with knowledge of well-accepted auditing processes of information systems, business resilience and IT operations. It also offers excellent insights into information assets protection, management, IT governance, development, implementation and information systems acquisition.

A candidate interested in acquiring the CISA certification must comply with the organization's requirements, including the information system auditing standards. A candidate must adhere to the code of ethics as well as the continuing education policy. In addition to passing the exam, a candidate must have at least five years of information systems auditing, assurance, control or security work experience within the past ten years of the application submission date. The experience should be aligned with the prescribed CISA practice domain areas.

According to ISACA, there are various experience waivers available to a candidate: a one-year waiver for an associate degree; a two-year waiver for a bachelor's, master's or doctorate in any field; or a three-year waiver for a master's degree in information systems or related field. There are also two-year waivers for a candidate who possesses the Chartered Institute of Management Accountants (CIMA) qualification and member status from the Association of Chartered Certified Accountants (ACCA).

To maintain the CISA certification, a candidate must attain the required Continuing professional education (CPE) hours, which is a minimum of 120 hours over three years, with at least 20 CPE hours every year.

For more information, go to https://www.isaca.org/credentialing/cisa.

CISM

The Certified Information Security Manager (CISM) certification is specifically established to assert a security professional's capabilities in security governance, security program development, incident management and enterprise risk management. A CISM designation demonstrates to employers a candidate's expertise in effectively leading an enterprise security program aligned to the business strategy. The CISM certification enables a professional to transition from a technology-focused role to a more management-focused one.

A candidate interested in acquiring the CISM certification must comply with its requirements, including adherence to ISACA's code of professional ethics and the continuing professional education (CPE) policy. In addition to passing the exam, a candidate must have at least five years of information security management work experience within the past ten years of the application submission date.

The experience should be aligned with the prescribed CISM practice domain areas—three out of the four domains. According to ISACA, there are various experience waivers available to a candidate: a one-year waiver for a bachelor's degree in information security or related field; a one-year waiver for a skill-based general security certification; a one-year waiver for information systems management experience; a two-year waiver for an MBA or a master's degree in information security or related field; and a two-year waiver for a CISA or CISSP in good standing.

To maintain the CISM certification, a candidate must attain the required continuing professional education (CPE) hours, which is a minimum of 120 hours over three years, with at least 20 CPE hours every year.

For more information, go to https://www.isaca.org/credentialing/cism.

CRISC

The Certified in Risk and Information Systems Control (CRISC) certification asserts a professional's ability and experience to identify and manage enterprise IT risks while implementing and maintaining information systems controls. The CRISC certification acts as a career catalyst as it increases a professional's credibility through instant recognition. A professional with the CRISC certification can assert a robust foundational understanding of information security risk concepts and their overall impact on the organization's operations and reputation. Furthermore, a professional can assert their ability to devise strategies and plans for managing potential risks within an enterprise's threshold. A CRISC professional also establishes the knowledge of relevant risk concepts, which empowers him or her to communicate efficiently within a group of IT professionals and business stakeholders.

A candidate interested in acquiring the CRISC certification must comply with its requirements, including adherence to ISACA's code of professional ethics and the continuing professional education (CPE) policy. In addition to passing the exam, a candidate must have at least three years of risk management and information security control experience within the past ten years of the application submission date. The experience should be aligned with the prescribed CRISC practice domain areas—two out of the four domains. Out of the two, one year should be in either IT risk identification or IT risk assessment domains. According to ISACA, there are no substitutions or experience waivers for the CRISC credential.

To maintain the CRISC certification, a candidate must attain the required continuing professional education (CPE) hours, which is a minimum of 120 hours over three years, with at least 20 CPE hours every year.

For more information, go to https://www.isaca.org/credentialing/crisc.

CGEIT

The Certified in Governance of Enterprise IT (CGEIT) is a vendor-neutral and framework-neutral certification demonstrating a professional's proficiency in advisory, assurance and IT governance knowledge areas. With businesses increasingly relying on information technology for growth, managing

IT can often lead to an actual or perceived conflict of interest between various stakeholders unless appropriate oversight and visibility is established. A CGEIT-certified professional helps establish proper operating models, reporting structures as well as roles and responsibilities to provide to manage risks related to IT.

A candidate interested in acquiring the CGEIT certification must comply with its requirements, including adherence to ISACA's code of professional ethics and the continuing professional education (CPE) policy. In addition to passing the exam, a candidate must have at least five years of experience managing or supporting IT governance within the past ten years of the application submission date. Out of the five years, one year should be directly in Domain 1 of the CGEIT (Governance of Enterprise IT). According to ISACA, there are no substitutions or experience waivers for the CGEIT credential.

To maintain the CGEIT certification, a candidate must attain the required continuing professional education (CPE) hours, which is a minimum of 120 hours over three years, with at least 20 CPE hours every year.

For more information, go to https://www.isaca.org/credentialing/cgeit.

CSX-P

The Cybersecurity Practitioner (CSX-P) is a certification demonstrating the abilities and skills of cybersecurity professionals mapped to the five security functions of the NIST Cyber Security Framework: Identify, Protect, Detect, Respond and Recover. The CSX-P is a performance-based, vendor-neutral certification that validates a cybersecurity professional's technical abilities and skills in a live virtual environment, similar to what a cyber incident first responder would encounter. The key knowledge areas covered in the CSX-P exam assess a candidate's analytical ability to identify and resolve real-life threat scenarios.

A candidate interested in acquiring the CSX-P certification must comply with its requirements, including adherence to ISACA's code of professional ethics and the continuing professional education (CPE) policy. To maintain the CSX-P certification, a candidate must attain the required continuing professional education (CPE) hours, which is a minimum of 120 hours over three years, with at least 20 CPE hours every year. ISACA outlines no other professional experience requirements for the CSX-P.

For more information, go to https://www.isaca.org/credentialing/csx-p.

CDPSE

The Certified Data Privacy Solutions Engineer (CDPSE) certification is a relatively new addition to ISACA's popular certifications. This certification validates a cybersecurity professional's ability to apply leading privacy practices within organizational platforms and products resulting in increased customer trust. With increasingly stringent privacy and data protection regulatory requirements, organizations require experienced professionals with validated credentials to help navigate the ensuing compliance complexities. A professional with the CDPSE certification can provide advisory services (based on validated experience) to stakeholders on data life cycle controls and policies, enabling organizations to consider privacy implications within business processes. The CDPSE certification asserts a candidate's expertise in the three knowledge areas of privacy governance, privacy architecture and data life cycle.

At the time of this publication, the CDPSE exam is in beta mode and should be released sometime in 2021. In the interim, a candidate interested in the credential can either take the beta test or apply for certification under ISACA's early adoption program by demonstrating five years' experience in at least two out of three knowledge areas. Two out of five years can be waived if a candidate has the CISA, CISM, CGEIT, CRISC, CSX-P or FIP certifications. Additionally, the candidate must accept and adhere to ISACA's code of professional ethics and the continuing professional education (CPE) policy.

To maintain the CDPSE certification, a candidate must attain the required continuing professional education (CPE) hours, which is a minimum of 120 hours over three years, with at least 20 CPE hours every year.

For more information, go to https://www.isaca.org/credentialing/certified-data-privacy-solutions-engineer.

EC-COUNCIL

The International Council of Electronic Commerce Consultants (EC-Council) is an international organization that strives to provide a diverse certification platform to cybersecurity professionals. The EC-Council has gained immense popularity due to its valuable certifications in IT that empower professional skills. The organization provides a significant number of certifications in specialties such as digital forensics, disaster recovery, IT security and software security.

The changing dynamics of the cybersecurity market tend to create a gap between the latest approach and IT practitioners. The EC-Council actively works to bridge this gap through its InfoSec certifications. A cybersecurity professional who acquires any of these certifications is considered a well-informed professional who has a diverse approach to solving complex issues. This chapter briefly explores 16 of EC-Council certifications, including the CSCU, CND, CEH, ECSA, LPT-MASTER and more.

For more information about EC-Council, visit https://www.eccouncil.org/programs.

CSCU

The EC-Council Certified Secure Computer User (CSCU) certification acts as a validation that a professional is well versed in security and networking skills. The CSCU certification allows a professional to assert formal recognition of her or his knowledge regarding a variety of network and computer security threats such as credit card fraud, identity theft, viruses, backdoors, emails hoaxes, online banking phishing scams, online sex offenders, confidential information loss, social engineering and hacking attacks.

This certification also establishes that a candidate understands and can act against security uncertainties and has learned ways to mitigate unauthorized security exposure. Having this certification nicely complements other education in the field of cybersecurity and data protection. The CSCU certification was established for today's connected computer users who uses the internet extensively in their daily lives.

There are certain legalities and restrictions when it comes to applying for CSCU certification. For example, the applicant must be 13 years old when applying for this certification. A candidate considering the CSCU certification exam must comply with the EC-Council examination policies.

For more information, go to https://www.ec-council.org/programs/certified-secure-computer-user-cscu.

CND

The Certified Network Defender (CND) certification was created predominantly based on a thorough job task analysis as represented in the NIST NICE Framework. The program also maps to DoD roles for system and network administrators. These network administrators work to detect, protect, respond and (where possible) predict network anomalies that might be indicators of attack. The majority of the CND program is based on hands-on practical skills created by EC-Council labs.

A CND-certified professional can substantiate a proper understanding of traffic, utilization, performance, network topology, network components, security policy and patterns based on system and user location. A CND-certified professional can demonstrate a fundamental understanding of network technologies, data transfer and software technologies. This, in turn, aids the professional to develop risk-based strategies that incorporate establishing resilient network perimeters and align with leading cyber defense practices.

The CND credential can be instrumental for a professional who seeks to further his or her career in the various cyber defense roles, such as security operations analyst, network security administrator or cybersecurity architects. For instance, a CND-certified network administrator can outline effective network security procedures and corresponding incident response strategies.

The EC-Council CND Certification program imparts practical knowledge about network security controls application, network defense fundamentals, perimeter appliances, protocols, secure VPN firewall configuration, Intrusion Detection System (IDS), network traffic signature intricacies, vulnerability scanning and analysis.

To obtain a CND certification, the candidate does not require any prescribed educational qualifications. However, EC-Council provides two options for a candidate considering taking the certification exam. This includes participation in the EC-Council's network security training or two years of information security work experience and submitting an eligibility application form.

For more information, go to https://www.ec-council.org/programs/certified-network-defender-cnd.

CEH

The advanced-level Certified Ethical Hacker (CEH) certification is one of the most valuable offensive security credentials as it substantiates an information security professional's existing ethical hacking skills. The CEH certification focuses explicitly on the technical aspect of identifying and attempting to exploit vulnerabilities within a system in a highly controlled and lawful environment.

System configurations and organizational assets are different in organizations, depending on their business and geography. Therefore, not all vulnerabilities present exploitable risks. To that end, professionals with hands-on expertise and the ability to "think like a hacker" contributes to a fully-rounded cyber defense team. A CEH credential can validate a professional's ability to assure that IT systems can withstand anomalous actors by identifying and fixing exploitable vulnerabilities before a threat actor can.

To obtain a CEH certification, the candidate does not require any prescribed educational qualifications. However, the EC-Council provides the following eligibility considerations when attempting the CEH certification examination: 1) to participate in the EC-Council's CEH training or 2) possess two years of information security work experience and submit an eligibility application form.

For more information, go to https://www.ec-council.org/programs/certified-ethical-hacker-ceh.

ECSA

The EC-Council Certified Security Analyst (ESCA) certification is a step ahead of the CEH certification. It involves completing a rigorous 12-hour practical exam that tests a professional's penetration testing skills in a real-world simulated environment. This environment closely replicates an organization with a segregated network infrastructure containing various database servers and other network devices simulating a real-world IT infrastructure.

From network scans to extensive vulnerability analysis, a candidate must demonstrate her or his offensive security and penetration testing skills to perform a comprehensive security audit of the system. From a hiring manager's point of view, achieving the ECSA certification means the candidate has effectively demonstrated a high degree of penetration testing capabilities and can successfully conduct various offensive security activities.

The ECSA certification has no set requirements when it comes to education. However, each applicant must be over 18 years of age. A candidate must have two years of working experience within the ethical hacking domain. In case a candidate does not have sufficient experience, a candidate must enroll in a professional training session as a waiver. Each candidate must comply with EC-Council's stated rules and regulations.

For more information, go to https://www.eccouncil.org/programs/certified-security-analyst-ecsa.

CPENT / LPT-MASTER

The EC-Council Certified Penetration Testing Professional (CPENT) and the Licensed Penetration Tester - Master (LPT-MASTER) is an advanced-level certification adopted by professionals who want to take their penetration testing expertise to the next level. This certification is challenging to acquire. It tests an offensive security professional's ability to test some of the most hardened systems while scoring at least 90% on the 24-hour exam. Candidates who cannot achieve 90% can still accomplish the CPENT certification if they can score higher than 70%. A professional can effectively demonstrate her or his master-level penetration testing skills upon achieving this certification.

The CPENT/LPT exam involves a "capture the flag" (CTF) scenario where professional ethical hackers attempt to solve complex cybersecurity problems with a high degree of difficulty. The CTF comprises a multi-layered network architecture, including requirements to circumvent web applications by carefully selecting techniques and tools required to exfiltrate data needed to complete the challenge.

In this certification program, a cybersecurity professional learns professional penetration testing and security skills. This program is designed by integrating concepts such as defense scanning, deploying proxy chains, pivoting between networks, and using web shells. The EC-Council proctors offer a watchful eye during the examination.

There are no predefined criteria for the CPENT/LPT-MASTER certification in terms of experience or education. However, an individual applying for CPENT/LPT-MASTER certification must be at least 18 years of age.

For more information, go to https://www.ec-council.org/programs/licensed-penetration-tester-lpt-master.

EISM

The EC-Council Information Security Manager (EISM) certification is a stepping stone toward the Certified Chief Information Security Officer (CCISO) designation. It is designed for individuals who do not yet have the five years of experience required for the latter but still want to pursue a management-centric career in information security. The EISM candidate must participate in the prescribed EC-Council training, which is the same training as security executives attend. The EISM program was developed for professionals striving to assert their understanding of the basics of information security management and establish a management-specific path for their careers.

Further details about the certification are available at https://ciso.eccouncil.org/cciso-certification/eism-program.

CCISO

The EC-Council Certified Chief Information Security Officer (CCISO) training and certification program is unique in that it targets the cybersecurity executive. There are three paths to obtaining the CCISO designation: self-study, training and by way of the EISM program (discussed above). Self-study candidates have the required information security management experience and are ready to write the exam. Candidates in this group also can apply the credit toward the experience requirements if they have specific professional certifications and higher degrees in information security.

The training option is open to anyone who does not meet the self-study option aforementioned. However, in addition to obtaining the training, candidates also need to demonstrate five years of IS management experience in three out of the five (3 out of 5) CCISO domains. The last path to CCISO certification is also by way of obtaining the EISM program, as discussed in the previous section. The CCISO certification is a powerful program that concentrates on the management aspect of information security through an executive's eyes.

For more information, go to https://ciso.ec-council.org/cciso-certification.

ECIH

The EC-Council Certified Incident handler (ECIH) program is designed to impart critical knowledge to cyber incident handlers, including various intricacies in handling post-breach activities. This certification is beneficial for professionals who want to demonstrate their expertise as incident responders or handlers. The training program developed by EC-Council for the ECIH certification includes hands-on learning delivered through a lab setting, including steps involved in planning, recovering and reporting.

This certification allows the individual to upsurge her or his abilities by acquiring the best methodologies within the domain. The ECIH certification enables a professional to analyze complex situations and evaluate a quick response systematically. It also empowers an organization to determine, restrict, mitigate and recover from an uncertain situation. The certification helps with drafting effective incident management policies and playbooks that form a critical part of an organization's Computer Security Incident Response Team (CSIRT).

To obtain the ECIH certification, a professional must comply with the EC-Council requirements, such as possessing one year of work experience in information security. This requirement can be removed if the candidate attends the official EC-Council ECIH training. The EC-Council must approve a candidate's eligibility application process to bypass the training requirement.

Further details about the certification are available at https://cert.eccouncil.org/ec-council-certified-incident-handler.html.

CHFI

The Computer Hacking Forensic Investigation (CHFI) is a highly reputable certification in the forensic investigation field. This certification validates a security professional's ability to utilize tools, methods and skills to analyze a cyberattack, extract proper evidence, establish a crime report and conduct audits to reduce such incidents from recurring in the future.

Global organizations and legal agencies alike need CHFI-certified employees to help carry out proper investigations. A certified CHFI employee can quickly determine the appropriate methodologies for recovering encrypted, deleted and damaged data during the data recovery process. The CHFI credential also validates the candidate's ability to look at an intruder's modus operandi through a fine-tooth comb and to gather the required evidence for the court of law.

To obtain the CHFI certification, a professional must comply with the EC-Council regulations, such as having two years of work experience in information security; this requirement can be removed if the candidate attends the official EC-Council CHFI training. The EC-Council must approve a candidate's eligibility application process to bypass the training requirement.

For more information, go to https://www.ec-council.org/programs/computer-hacking-forensic-investigator-chfi.

EDRP

During uncertain times and unforeseen disasters, such as pandemics or natural calamities, many companies now appreciate the value of a well-tested IT recovery and availability program. The EC-Council Disaster Recovery Professional (EDRP) certification provides a professional with a comprehensive understanding of disaster recovery and business continuity principles, including carrying out risk assessments and business impact analyses to establish recovery priorities. The EDRP-certified professional helps businesses ensure IT environments can be recovered based on prioritizing data and assets established in close coordination with business groups and IT partners.

To obtain the EDRP certification, a professional must comply with the EC-Council regulations, such as having two years of work experience in information security. This requirement can be removed if the candidate attends the official EC-Council EDRP training or by purchasing an official courseware bundle directly from EC-Council. The EC-Council must approve a candidate's eligibility application process to bypass the training requirement.

For more information, go to https://www.ec-council.org/programs/disaster-recovery-professional-edrp.

ECES

The EC-Council Certified Encryption Specialist (ECES) certification is another increasingly popular certification. It is a one-of-a-kind program explicitly geared toward the highly skilled field of cryptography and the science behind encryption. A professional establishes a comprehensive understanding of crucial cryptography and its essential dynamics through this certification.

The ECES certification also delves into further details of cryptographic algorithms such as AES, DES and Feistel networks. The ECES certification provides an overview of the cryptography algorithm, hashing algorithm, asymmetric cryptography, and other significant concepts such as diffusion and confusion principles. A candidate also acquires practical skills such as setting up a Virtual Private Network (VPN), encrypting a drive, steganography experience and experience with cryptography algorithms.

To obtain the ECES certification, a professional must comply with the EC-Council regulations, such as one year of work experience in information security. This requirement can be removed if the candidate attends the official EC-Council ECES training or by purchasing an official courseware bundle directly from EC-Council. The EC-Council must approve a candidate's eligibility application process to bypass the training requirement.

For more information, go to https://www.ec-council.org/programs/ec-council-certified-encryption-specialist-eces.

CASE JAVA and CASE .NET

The Certified Application Security Engineer (CASE) certification is established with global software development experts' collaboration. The certification training program consists of security strategies integrated at various phases of the Software Development Lifecycle (SDLC). A security professional with CASE JAVA and CASE.NET certification can champion web-facing software applications with security in the DNA. Each of the CASE certifications (JAVA and .NET) provides a factual assertion about the candidate's understanding of ensuring security is a part of the entire code life cycle, from requirements to maintenance.

To obtain the CASE JAVA or CASE.NET certifications, a professional must comply with the EC-Council regulations, such as having two years of work experience in information security. This requirement can be removed if the candidate attends the official EC-Council CASE pieces of training or by purchasing an official courseware bundle directly from EC-Council. The EC-Council must approve a candidate's eligibility application process to bypass the training requirement.

For more information, go to https://www.ec-council.org/programs/certified-application-security-engineer-case.

CTIA

Incorporating threat intelligence data feeds into an organization's cyber defense capabilities is one of the most crucial aspects of standing up a resilient business environment. In that light, the Certified Threat Intelligence Analyst (CTIA) credential asserts an individual's capability to help organizations be proactive in handling and thwarting today's advanced persistent threats by proactively investigating and benchmarking anomalies, which could be indicators of compromise. There is an increased demand for trained individuals who can build strategies for extracting intelligence from data by implementing advanced capabilities using various relevant channels.

To obtain the CTIA certification, a professional must comply with the EC-Council regulations, such as possessing two years of work experience in information security. This requirement can be removed if the candidate attends the official EC-Council CTIA training. The EC-Council must approve a candidate's eligibility application process to bypass the training requirement..

For more information, go to https://www.ec-council.org/programs/certified-threat-intelligence-analyst-ctia.

CSA

The Certified SOC Analyst (CSA) certification applies to the various skills required to join a Security Operations Center (SOC). It is crafted for aspiring and recent Tier I and II SOC analysts to perform relevant cyber defense tasks. The CSA certification is a credentialing and training program that assists a professional in acquiring relevant skills to provide high-quality security services, such as detecting and countering potential cyber threats.

The CSA certification program covers all relevant topics applicable to SOC operations within a three-day program. This certification asserts that an individual has the appropriate knowledge for correlation, log management, advanced incident detection, incident response and SIEM deployment. The CSA training program also covers various SOC processes and tools required to effectively collaborate with the Computer Security Incident Response Team (CSIRT) when needed.

To obtain the CSA certification, a professional must comply with the EC-Council regulations, including having one year of work experience in information security. This requirement can be removed if the candidate attends the official EC-Council CSA training or by purchasing an official courseware bundle directly from EC-Council. The EC-Council must approve a candidate's eligibility application process to bypass the training requirement.

Further details about the certification can be found at https://www.eccouncil.org/wp-content/up-loads/2019/08/CSA-Brochure.pdf.

ECSS

The EC-Council Certified Security Specialist (ECSS) certification provides a firm footing for an individual to improve his or her abilities in various information security facets, such as network security and computer forensics. This certification is an entry-level program covering the foundational aspect of information security. The training program is built to acquaint an individual with the various foundational aspects of information security.

To obtain the ECSS certification, a professional must comply with the EC-Council regulations, such as possessing one year of work experience in information security. This requirement can be removed if the candidate attends the official EC-Council CSA training or by purchasing an official courseware bundle directly from EC-Council. The EC-Council must approve a candidate's eligibility application process to bypass the training requirement.

For more information, go to https://www.ec-council.org/programs/certified-security-specialist-ecss.

OFFENSIVE SECURITY

Offensive Security, also known as OffSec, started in 2007 to provide advanced security courses and open source projects. One of their most notable open-source platforms is the Kali Linux distribution, a globally recognized platform used for penetration testing. The organization also provides training programs for entry-level professionals who may struggle to progress within their careers. OffSec offers various professional certifications that assist in enhancing the abilities of cybersecurity professionals. The organization currently provides the following certifications, and each certification focuses on a different security domain:

- Offensive Security Certified Professional (OSCP)
- Offensive Security Wireless Professional (OSWP)
- Offensive Security Experienced Penetration Tester (OSEP)
- Offensive Security Web Expert (OSWE)
- Offensive Security Exploit Developer (OSED)
- Offensive Security Exploitation Expert (OSEE)
- Offensive Security Certified Expert (OSCE)

For more information about OffSec, visit https://www.offensive-security.com.

OSCP

The Offensive Security Certified Professional (OSCP) certification is a technical certification focused on a professional's ability to detect system vulnerabilities. It helps develop skill sets required for a professional to determine current and exploitable vulnerabilities while also learning how to implement organized strategies to address the vulnerabilities.

The OSCP certification is widely respected and generally required for various popular cybersecurity positions, including penetration testers, malware specialists, security analysts and threat hunters.

A candidate seeking to achieve the OSCP certification must fulfill the set requirements and prove her or his eligibility according to the set standards. The candidate must have a strong understanding of IP and TCP networking and practical experience with Linux and Windows administration. She or he should also know about script bashing with programming languages such as Perl or basic Python. The OSCP exam consists of a hands-on penetration test within an isolated network that lasts for up to 24 hours.

Further details about the certification are available at https://www.offensive-security.com/pwk-oscp.

OSWP

The Offensive Security Wireless Professional (OSWP) certification is established specifically for a network security professional who seeks to assess the capability of identifying vulnerabilities and auditing 802.11 wireless networks. A candidate with OSWP certification knows how to induce controlled attacks to determine the potential uncertainties within a network. An OSWP certified professional can gather wireless information, access restrictions within a circumvented wireless network and crack wireless encryptions such as WPA, WPA2, WEP as well as execute man-in-the-middle attacks.

To acquire the OSWP certification, a candidate must have the predefined knowledge required to appear in the exam. A candidate appearing in the exam must comprehend the OSI model and IP/TCP networking. The exam lasts for four hours, in which a candidate has to submit a comprehensive penetration test report.

For more information, go to https://www.offensive-security.com/wifu-oswp.

OSEP

The Offensive Security Experienced Penetration Tester (OSEP) is an advanced penetration testing certification. It is recommended for security professionals to first obtain the OSCP before attaining this certification. The OSEP certification tests a candidate's understanding of evasion techniques, breaching configured defenses and exploiting and penetrating hardened systems.

To acquire the OSEP certification, a candidate must have a working knowledge of Kali Linux; programming languages such as C++; and scripting capabilities in Bash, Python and PowerShell. The exam is conducted and administered over a 48-hour period.

For more information, go to https://www.offensive-security.com/pen300-osep.

OSWE

The Offensive Security Web Expert (OSWE) certification is an advanced-level web application security certification. It is suggested a candidate should attain the OSCP certification before applying for the OSWE certification. A certified OSWE professional has impeccable knowledge about decompiled codes, identification of logical vulnerabilities and vulnerability exploitation. The OSWE certification provides a practical and clear understanding of the web application hacking process and application assessment. This certification assesses the ability of a candidate to identify vulnerabilities, code web apps and exploit vulnerabilities.

A candidate appearing for the OSWE examination must be familiar with operating systems such as Windows and Linux; and capable of writing programs and shell scripts using PHP, Python, Bash and Perl; and have a strong knowledge of web proxies, web attack theory, vector and practices. The OSWE examination lasts for up to 48-hours and includes a hands-on web application assessment.

For more information, go to https://www.offensive-security.com/awae-oswe.

OSED

The Offensive Security Exploit Developer (OSED) certification is an intermediate exploit development certification. The OSED certification tests a candidate's exploit development skills and provides her or him with an overall understanding of exploiting vulnerabilities within systems.

To acquire the OSED certification, a candidate must be familiar with debuggers; writing code in languages such as Python; can read and understand C-code and 32-bit assembly at a basic level. The exam is conducted and administered over a 48- hour period.

For more information, go to https://www.offensive-security.com/exp301-osed.

OSEE

The Offensive Security Exploitation Expert (OSEE) certification is designed to build complex exploitation abilities within a professional. It is advised to enroll and achieve the OSCE certification before applying for OSEE certification. The OSEE is a hands-on penetrating testing certification that validates a candidate's ability to navigate advanced vulnerabilities, adapts to emerging technical security challenges, finds incorrect codes, analyzes vulnerable software and develops functional exploits effectively and efficiently.

A candidate interested in acquiring the OSEE certification must have ample knowledge regarding exploits and know ways to operate tools for debugging, scripting and programming. The OSEE examination is 72-hours in which a candidate must create strong strategies and exploits to protect and determine vulnerabilities within an environment then submit a comprehensive penetration test report as part of the exam.

For more information, go to https://www.offensive-security.com/awe-osee.

OSCE

The Offensive Security Certified Expert (OSCE) is a certification that deeply analyzes a cybersecurity professional's penetration testing skills. This certification is a self-paced course for ethical hacking where a candidate learns how to determine misconfigurations and advanced vulnerabilities in diverse operating systems. The OSCE certification enables a cybersecurity professional to plan controlled attacks, implement an organized attack on weak systems and acquire administrative access. The OSCE certification is a prerequisite for the Offensive Security Exploitation Expert (OSEE) certification.

A candidate applying for the OSCE certification exam must have a solid knowledge of penetration testing to perform well in the exam. A candidate should have a comprehensive understanding of shellcode encoding concepts, Metasploit frameworks and operating systems such as Windows and Linux. The OSCE exam has a 48-hour time limit with a comprehensive test report as part of the exam.

For more information, go to https://www.offensive-security.com/ctp-osce.

GIAC

Founded in 1999 by the SANS Institute, the Global Information Assurance Certification (GIAC) provides several technical cybersecurity certifications. GIAC certifications are considered premium cybersecurity certifications and are globally well recognized. Currently, GIAC certifications are designed to guarantee proficiency in specialized, critical cybersecurity domains. These certifications are categorized into eight areas as follows:

	Category	# of Certifications	Certifications
1	Cyber Defense	13	GISF, GSEC, GOSI, GCED, GPPA, GCIA, GCWN, GCUX, GMON, GDSA, GCDA, GCCC, GDAT
2	Industrial Control Systems (ICS)	3	GICSP, GRID, GCIP
3	Offensive Operations	9	GCIH, GEVA, GPEN, GWAPT, GPYC, GMOB, GCPN, GAWN, GXPN
4	Digital Forensics and Incident Response (DFIR)	7	GCFE, GBFA, GCFA, GNFA, GCTI, GASF, GREM
5	Cloud Security	3	GWEB, GCSA, GCLD
6	Management and Leadership	6	GISP, GSLC, GSTRT, GCPM, GLEG, GSNA
7	GIAC Security Expert	1	GSE
8	Other GIAC Certifications	2	GSSP-JAVA, GSSP-.NET

The GIAC organization distinguishes itself in providing employers assurance that prospective and existing cybersecurity practitioners are knowledgeable, skilled and can perform their roles well. In this section, we will briefly describe the GIAC certifications.

Further details about GIAC are available at https://www.giac.org.

GIAC Certifications: Cyber Defense Category

GISF

The GIAC Information Security Fundamentals (GISF) is an introductory-level certification that assesses a professional's knowledge of computer functions, security foundation, networking, cybersecurity technologies and introductory level cryptography. The GISF certification requires a candidate to complete one proctored exam with a passing threshold of 72%.

For more information, go to https://www.giac.org/certification/information-security-fundamentals-gisf.

GSEC

The GIAC Security Essentials (GSEC) certification is designed to technically challenge an information security practitioner to analyze his or her abilities apart from simple concepts and terminology. The GSEC certification holder can execute all the information security responsibilities by adopting the latest technological approaches.

A certified GSEC professional can establish an active defense mechanism through effective password management and access control. A GSEC professional has ample knowledge regarding cryptography algorithms, applications and deployment. The GSEC certification validates a professional's ability in domains, such as defensible network architecture, incident response, incident handling, Linux security, Windows, web application and security policies. The GSEC certification requires a candidate to complete one proctored exam with a passing threshold of 73%.

For more information, go to https://www.giac.org/certification/security-essentials-gsec.

GOSI

The GIAC Open Source Intelligence (GOSI) is an advanced-level certification that assesses a professional's knowledge in open source intelligence methodologies and frameworks, open-source intelligence data collection, analysis, reporting as well as harvesting data from the dark web. The GOSI certification requires a candidate to complete one proctored exam with a passing threshold of 76%.

For more information, go to https://www.giac.org/certification/open-source-intelligence-gosi.

GCED

The GIAC Certified Enterprise Defender (GCED) is an advanced-level certification that builds on the GIAC Security Essentials certification. It validates more technical, advanced abilities required to protect an organization's environment and defend an organization overall.

The certified GCED professional is highly knowledgeable about packet analysis, defensive network infrastructure, penetration testing, malware removal, incident handling, computer crime investigation, hacker tools, and network and computer hacker exploits. The GCED certification requires a candidate to complete one proctored exam with a passing threshold of 70%.

For more information, go to https://cyber-defense.sans.org/certification/gced.

GPPA

The GIAC Certified Perimeter Protection Analyst (GPPA) is an advanced-level certification that evaluates a professional's ability to design, monitor, and configure routers and firewalls. A certified GPPA professional has comprehensive skills and knowledge to configure, monitor and frame perimeter defense systems. A GPPA certificate holder can implement, monitor, and secure design perimeters for an organization, including firewalls, routers, overall network design, VPNs, and remote access. The GPPA certification requires a candidate to complete one proctored exam with a passing threshold of 69.3%.

For more information, go to https://www.giac.org/certification/certified-perimeter-protection-analyst-gppa.

GCIA

The GIAC Intrusion Analyst (GCIA) is an advanced-level certification that accesses knowledge of host monitoring, intrusion detection and traffic analysis. The certified GCIA professional acquires the skills required to monitor and configure intrusion detection systems as well as acquires the ability to interpret, analyze and read related log files and network traffic. The GCIA certification provides a firm ground for professionals who understand the fundamentals of application protocol and traffic analysis, open-source IDS, network traffic monitoring and forensics. The GCIA certification requires a candidate to complete one proctored exam with a passing threshold of 68%.

For more information, go to https://www.giac.org/certification/certified-intrusion-analyst-gcia.

GCWN

The GIAC Certified Windows System Administrator (GCWN) is an advanced-level certification that assesses a professional's ability to protect Microsoft Windows servers and clients. The certified GCWN professional acquires the skills and knowledge required to manage and configure the security of operating applications and systems.

A professional can protect Microsoft operating applications and systems such as IPsec, PKI, AppLocker, DNSSEC, Group Policy, hardening, as well as PowerShell Windows against the persistent adversary and malware. The GCWN certification covers defensible networking, endpoint protection, PKI management, securing PowerShell, restricting administrative compromise and application hardening. The GCWN certification requires a candidate to complete one proctored exam with a passing threshold of 66%.

For more information, go to https://www.giac.org/certification/certified-windows-security-administrator-gcwn.

GCUX

The GIAC Certified Unix System Administrator (GCUX) is an advanced-level certification that validates the skills and knowledge of a professional in areas such as Linux application security, Linux and Unix digital forensics and hardening Linux and Unix systems. A certified GCUX professional has the abilities, skills and knowledge to audit and protect Linux and UNIX systems. This professional can also utilize a variety of tools to combat security issues, including AIDE, Sudo, lsof, SSH and many others. The GCUX certification requires a candidate to complete one proctored exam with a passing threshold of 68%.

For more information, go to https://www.giac.org/certification/certified-unix-security-administrator-gcux.

GMON

The GIAC Continuous Monitoring (GMON) is an advanced-level certification that assesses an information security professional's expertise to determine uncertainties and rapidly identify suspicious activity. A certified GMON professional has viable knowledge about defensible security network security monitoring, architecture, continuous mitigation and diagnostic and continuous security monitoring.

Through GMON certification, a candidate elevates his or her knowledge in areas such as the security operation center (SOC) and SOC security architecture, network security architecture, network security monitoring, endpoint security architecture, continuous monitoring and automation. The GMON certification requires a candidate to complete one proctored exam with a passing threshold of 74%.

For more information, go to https://www.giac.org/certification/continuous-monitoring-certification-gmon.

GDSA

The GIAC Defensible Security Architecture (GDSA) is an advanced-level certification that assesses a professional's knowledge of defensible security architecture, network security architecture and zero trust architecture. The GDSA certification is designed for security architects, security engineers, system administrators, cyber threat investigators and more. The GDSA requires a candidate to complete one proctored exam with a passing threshold of 63%.

For more information, go to https://www.giac.org/certification/defensible-security-architecture-gdsa.

GCDA

The GIAC Certified Detection Analyst (GCDA) is an advanced industry-level certification that evaluates a professional's ability to analyze, collect and utilize endpoint data and modern network sources to investigate the unauthorized or malicious activity. The GCDA certification validates a candidate's knowledge and ability to use wield tools such as Security Information and Event Management (SIEM). It demonstrates his or her abilities to utilize the specified tools to transform a threat actor's strengths into a threat actor's weaknesses.

A certified GCDA professional can benefit the organization by using polished skills such as advanced endpoint analytics, service profiling, user behavior monitoring and baselining. The GCDS professional also has sufficient knowledge regarding SIEM architecture, tactical SIEM detection, SOF-ELK and post-mortem analysis. The GCDA certification requires a candidate to complete one proctored exam with a passing threshold of 79%.

For more information, go to https://www.giac.org/certification/certified-detection-analyst-gcda.

GCCC

The GIAC Critical Controls Certification (GCCC) is an advanced-level certification focused on critical security control. The GCCC certification is the only certification that is based upon a risk-based, prioritized security approach. The GCCC certification validates that a candidate has the skills and knowledge to execute and implement the critical security controls suggested by the cybersecurity council and implement audits based on the standard.

The GCCC certification integrates knowledge within a candidate regarding Center for Internet Security (CIS) control, application software security, account monitoring, moderate utilization of administrative privileges, boundary defense and need-to-know access. Through GCCC certification, a candidate also learns about data recovery capability, data protection, software and hardware assets inventory control, control and limitation of network ports, and email and web browser protections. The GCCC certification requires a candidate to complete one proctored exam with a passing threshold of 71%.

For more information, go to https://www.giac.org/certification/critical-controls-certification-gccc.

GDAT

The GIAC Defending Advanced Threat (GDAT) is a unique advanced-level certification covering both defensive and offensive security topics in detail. The certified GDAT professional has comprehensive knowledge about the various strategies and mechanisms the adversary uses to penetrate through networks and the security controls that can restrict the adversary effectively. A GDAT certified professional effectively determines the extent of an attack and can rapidly respond to an uncertain situation with adequate solutions. These skills combined increase a cyber professional's ability to restrict and deter Advanced Persistent Threats (APTs).

An organization with GDAT certified employees can detect and prevent exploitation, post-exploitation activities and payload deliveries. The organization further benefits by using cyber deception for threat hunting, adopting advanced persistent threat methodologies and models for adequate protection. The GDAT certification requires a candidate to complete one proctored exam with a passing threshold of 70%.

For more information, go to https://www.giac.org/certification/defending-advanced-threats-gdat.

GIAC CERTIFICATIONS: ICS CATEGORY

GICSP

The Global Industrial Cybersecurity Professional (GICSP) is an intermediate-level certification. This certification aims to bridge the extensive gap between engineering, cybersecurity and IT to frame solutions for industrial security systems. The GISP certification falls in the vendor-neutral category. This certification is a joint effort of GIAC and some high-profile organizations that deploy, operate, maintain and establish control system infrastructure and industrial automation. The GICSP certification examines the knowledge of professionals who support as well as engineer control systems. The GICSP certification requires a candidate to complete one proctored exam with a passing threshold of 71%.

For more information, go to https://www.giac.org/certification/global-industrial-cyber-security-professional-gicsp.

GRID

The GIAC Response and Industrial Defense (GRID) is an advanced-level certification. This certification validates a professional's skill in implementing and designing active defense policies appropriate and specific to an Industrial Control System (ICS) and associated network. A certified GRID professional has a comprehensive understanding of ICS-specific attacks, active defense strategies and methodologies to mitigate these attacks. The GRID certification integrates the approach and key techniques relevant to core ICS subjects such as digital forensics, incident response (DFIR) and network security monitoring (NSM). The GRID certification requires a candidate to complete one proctored exam with a passing threshold of 74%.

For more information, go to https://www.giac.org/certification/response-industrial-defense-grid

GCIP

The GIAC Critical Infrastructure Protection (GCIP) is an advanced-level certification in the ICS domain. This certification empowers a professional by validating his or her ability to support, maintain and access critical systems. The GRID certification helps a professional gain knowledge regarding the regulatory needs of NERC CIP and practical application approaches to adhere to both cybersecurity objectives and regulatory compliance. Through GCIP certification, a professional is well-aware of the nuances of CIP standard applicability, NERC and strategies, as well as identification of the BES Cyber System. The GCIP certification requires a candidate to complete one proctored exam with a passing threshold of 70%.

For more information, go to https://www.giac.org/certification/critical-infrastructure-protection-gcip.

GIAC Certifications: Offensive Operations Category

GCIH

The GIAC Incident Handler (GCIH) is an intermediate-level certification that validates a cybersecurity professional's skills and abilities to respond, resolve and detect computer security incidents. A certified GCIH professional has the in-depth knowledge required for security incident management.

A certified GCIH professional has a common understanding of attack vectors, tools and techniques needed to respond and defend against such uncertain situations when they arise. A GCIH certification provides a professional with the knowledge of hacking tools, computer crime investigation, incident handling, and network and computer hacker exploits. The GCIH certification requires a candidate to complete one proctored exam with a passing threshold of 70%.

For more information, go to https://www.giac.org/certification/certified-incident-handler-gcih.

GEVA

The GIAC Enterprise Vulnerability Assessor (GEVA) is an advanced-level certification that assesses a professional's knowledge of vulnerability assessment framework planning and methodology, discovery and validation of vulnerabilities, and remediation and reporting techniques utilizing proper data management. The GEVA certification requires a candidate to complete one proctored exam with a passing threshold of 71%.

For more information, go to https://www.giac.org/certification/enterprise-vulnerability-assessor-geva.

GPEN

The GIAC Penetration Tester (GPEN) is an advanced-level certification for penetration testing that demonstrates the ability of a professional to adequately carry out a penetration test using best practice methodologies and techniques. The certified GPEN professional has the skills and background knowledge to engage in reconnaissance and conduct exploits, as well as use a process-oriented strategy to penetrate testing activities. A certified GPEN professional conducts comprehensive pen test scoping and planning, scanning, pivoting and exploitation along with web application penetration testing. The GPEN certification requires a candidate to complete one proctored exam with a passing threshold of 75%.

For more information, go to https://www.giac.org/certification/penetration-tester-gpen.

GWAPT

The GIAC Web Application Penetration Tester (GWAPT) is an advanced-level certification that validates a professional has efficient knowledge of web application penetration testing and exploits methodology. Through the GWAPT application, a professional can polish SQL injection attacks, web application management and testing tools skills. She or he is also an expert in authentication attacks, configuration testing, cross-site request forgery, web application overview, reconnaissance and mapping, and client injection attacks. The GWAPT certification requires a candidate to complete one proctored exam with a passing threshold of 71%.

For more information, go to https://www.giac.org/certification/web-application-penetration-tester-gwapt.

GPYC

The GIAC Python Coder (GPYC) is an advanced-level certification that validates a professional's comprehension of core programming concepts and the skills to analyze and write working code by way of the Python programming language. The certified GPYC professional has ample knowledge of creating custom tools, Python libraries, interacting with databases and websites, automating testing and collecting information on a network. The GPYC certification requires a candidate to complete one proctored exam with a passing threshold of 67%.

For more information, go to https://www.giac.org/certification/python-coder-gpyc.

GMOB

The GIAC Mobile Device Security Analyst (GMOB) is an advanced-level certification that ensures a cybersecurity professional responsible for protecting networks and systems is well-aware of adequate methods to properly secure the digital devices connected to vital information. The certified GMOB professional has significant knowledge about managing and accessing application security and protecting a mobile device against malware attacks. The GMOB certification requires a candidate to complete one proctored exam with a passing threshold of 71%.

For more information, go to https://www.giac.org/certification/mobile-device-security-analyst-gmob.

GCPN

The GIAC Cloud Penetration Tester (GCPN) is an advanced-level certification that assesses a professional's knowledge of conducting penetration tests on the cloud. It tests a candidate's ability on cloud penetration testing fundamentals, cloud services from service providers such as AWS and Azure and potential cloud service attacks, and cloud-native applications with containers. The GCPN certification requires a candidate to complete one proctored exam with a passing threshold of 70%.

For more information, go to https://www.giac.org/certification/cloud-penetration-tester-gcpn.

GAWN

The GIAC Assessing Wireless Network (GAWN) is an advanced-level certification designed for a technology expert who constantly evaluates wireless network security. The GAWN certification concentrates on the diverse security aspects within the wireless network domain and the techniques and tools used to exploit wireless networks. Candidates attempting the GAWN certification need to have a comprehensive understanding of wireless networking. The GAWN certification requires a candidate to complete one proctored exam with a passing threshold of 70%.

For more information, go to https://www.giac.org/certification/assessing-auditing-wireless-networks-gawn.

GXPN

The GIAC Exploit Researcher and Advanced Penetration Tester (GXPN) is an advanced-level certification that demonstrates the skills of a professional to detect and eliminate possible security issues from networks and systems. The certified GXPN professional has the skills to perform advanced-level penetration tests and outline the nature of the threat actor to enhance the security system. This professional also knows how to evaluate the business risk attached to the behaviors in question. Through GXPN certification, a candidate enhances his or her Scapy, Python and fuzzing skills while also determining network attacks and conducting network booting in a restricted environment. The GXPN certification requires a candidate to complete one proctored exam with a passing threshold of 67%.

For more information, go to https://www.giac.org/certification/exploit-researcher-advanced-penetration-tester-gxpn.

GIAC Certifications: Digital Forensics and Incident Response (DFIR) Category

GCFE

The GIAC Certified Forensic Examiner (GCFE) is an intermediate-level certification that demonstrates the ability of a professional to conduct computer forensic analysis. This certification focuses on the core competencies needed to analyze and collect information from Windows computer systems. The certified GCFE professional has the skills, abilities, and knowledge to carry out complex incident investigations, including computer and cloud storage forensic analysis, reporting, e-Discovery, evidence gathering, browser forensics, acquisition, tracing application activities and acquisition on Windows systems.

A certified GCFE professional analyzes artifacts such as shell items, email, event logs, and USB devices through the use of techniques and tools such as keyword search, Windows registry forensic tools and web browser forensic tools. The GCFE certification requires a candidate to complete one proctored exam with a passing threshold of 71%.

For more information, go to https://www.giac.org/certification/certified-forensic-examiner-gcfe.

GBFA

The GIAC Battlefield Forensics and Acquisition (GBFA) is an intermediate-level certification that assesses a professional's knowledge of how to effectively collect data from a wide range of forensics analysis devices to rapidly produce actionable intelligence; and how to manually identify and acquire data. The

GBFA certification requires a candidate to complete one proctored exam with a passing threshold of 69%.

For more information, go to https://www.giac.org/certification/battlefield-forensics-acquisition-gbfa.

GCFA

The GIAC Certified Forensic Analyst (GCFA) certification validates a professional's advanced-level skills, abilities and knowledge to carry out detailed incident investigations and manage advanced incident handling situations. These situations include external and internal advanced persistent threats, data breach intrusions, anti-forensic policies utilized by complex digital forensic and attacker cases.

The GCFA certification emphasizes establishing core skills needed to gather and examine data from Linux and Windows computer systems. A certified GCFA professional effectively conducts time analysis, anti-forensic detection, memory forensic, threat hunting and advanced forensic response. The GCFA certification requires a candidate to complete one proctored exam with a passing threshold of 72%.

For more information, go to https://www.giac.org/certification/certified-forensic-analyst-gcfa.

GNFA

The GIAC Network Forensic Analyst (GNFA) is an advanced-level certification that demonstrates a professional's ability to employ network forensic examination and artifact analysis. The certified GNFA professional has a strong comprehension of network forensics fundamentals, abnormal conditions for network protocols, and advanced proficiency in network analysis tools to analyze system logs, devices, encrypted protocols, and wireless communication. Furthermore, the GNFA certification validates a professional's knowledge of open-source network security, NetFlow analysis, network protocol, network architecture and more. The GCFE certification requires a candidate to complete one proctored exam with a passing threshold of 70%.

For more information, go to https://www.giac.org/certification/network-forensic-analyst-gnfa.

GCTI

The GIAC Threat Intelligence (GCTI) is an advanced-level certification that holds a lot of significance for professionals working in the cybersecurity intelligence domain. In the evolving landscape of cybersecurity, a certified GCTI professional plays a crucial role. The GCTI certification amplifies the technical skills of a professional and stimulates critical thinking skills. A certified GCTI professional has a strong knowledge of operational, tactical and strategic cyber threat intelligence fundamentals. A professional performs intrusion analysis, collect as well as store data sets and retain the integrity of an organization's security system. The GCTI certification requires a candidate to complete one proctored exam with a passing threshold of 71%.

For more information, go to https://www.giac.org/certification/cyber-threat-intelligence-gcti

GASF

The GIAC Advanced Smartphone Forensics (GASF) is an advanced-level certification tailored to a professional closely associated with mobile device intricacies. The GASF certification plays a crucial role today as organizations are increasingly moving toward mobile applications for easy flow of communication. A GASF professional is knowledgeable about mobile forensics fundamentals, mobile application behavior, event artifacts analysis and mobile device malware analysis. The GASF certification requires a candidate to complete one proctored exam with a passing threshold of 69%.

For more information, go to https://www.giac.org/certification/advanced-smartphone-forensics-gasf.

GREM

The GIAC Reverse Engineering Malware (GREM) is an advanced-level certification validating a professional's ability to protect an organization against malicious attacks. It is designed for system and network administrators, auditors, forensic investigators and more. A certified GREM professional accumulate the skills and knowledge to reverse engineer malicious devices or software, such as web browsers and Microsoft Windows. A GREM professional uses her or his knowledge to conduct formal incident response, Windows system administration and forensic investigations. The GREM certification integrates malicious analysis skills empowering a professional with the latest knowledge and techniques. The GREM certification requires a candidate to complete one proctored exam with a passing threshold of 70.7%.

For more information, go to https://www.giac.org/certification/reverse-engineering-malware-grem.

GIAC Certifications: Cloud Security Category

GWEB

The GIAC Web Application Defender (GWEB) is an advanced-level cloud certification. The GWEB certification allows a professional to master his or her security knowledge of how to tackle common web application issues that often lead to multiple security issues. A successful professional should be proficient in using tools to identify and limit cross-site scripting (XSS), SQL injection and validation flaws. The GWEB certification also validates a professional's skill sets with access control, session management, authentication and defensive best practices. A certified GWEB professional has the information, expertise and skills required to secure web applications as well as identify and eliminate security shortcomings in current web applications. The GWEB certification requires a candidate to complete one proctored exam with a passing threshold of 68%.

For more information, go to https://www.giac.org/certification/certified-web-application-defender-gweb.

GCSA

The GIAC Cloud Security Automation (GCSA) is an advanced-level cloud security certification covering modern DevSecOps practices and cloud services used to deploy and establish applications and systems more securely. The GCSA certification indicates that a candidate is knowledgeable about DevSecOps and modern cloud principles and can also put theory into practice in a repeatable and automated manner.

A certified GCSA professional can demonstrate continuous integration, automating configuration management, continuous monitoring and continuous delivery skills. A GCSA certified professional knows how to use open-source tools, Microsoft Azure services, and service tool chains of Amazon Web Services. The GCSA certification requires a candidate to complete one proctored exam with a passing threshold of 61%.

For more information, go to https://www.giac.org/certification/cloud-security-automation-gcsa.

GCLD

The GIAC Cloud Security Essentials (GCLD) is an advanced-level certification that assesses a professional's knowledge of the opportunities and challenges presented when migrating systems and applications to cloud service provider environments; this includes planning, deploying, hardening, and securing single and multi-cloud environments, basic cloud resource auditing, security assessment and incident response. A professional appearing for the GCLD certification exam must oblige all the terms and conditions stated on the official website.

For more information, go to https://www.giac.org/certification/cloud-security-essentials-gcld.

GIAC Certifications: Management and Leadership Category

GISP

The GIAC Information Security Professional (GISP) is an intermediate-level certification that validates a professional's knowledge and skills specific to the CISSP certification. A certified GISP professional has practical knowledge of network communications, network security, asset security, access and identity management, risk and security management, asset security, security testing and assessment, security operation, software development security and security engineering. The GISP certification requires a candidate to complete one proctored exam with a passing threshold of 70%.

For more information, go to https://www.giac.org/certification/information-security-professional-gisp.

GSLC

The GIAC Security Leadership (GSLC) is an advanced-level certification within the management and leadership domain. This certification validates a professional's understanding of technical controls and governance focused on detecting, responding and protecting against security issues. The certified GSLC professional has a proficient knowledge of network, data, application, user and cost controls, and fundamental management subjects catering to the entire security life cycle. A certified GSLC professional can effectively manage application security, structure programs, manage security operations, regulate security policy and business continuity. The GSLC certification requires a candidate to complete one proctored exam with a passing threshold of 65%.

For more information, go to https://www.giac.org/certification/security-leadership-gslc.

GSTRT

The GIAC Strategic Planning, Policy and Leadership (GSTRT) is an advanced-level certification in the management and leadership domain. The GSTRT certification ensures that a professional has a comprehensive understanding of maintaining and developing cybersecurity programs. This certification also validates a professional's skill sets specific to business analysis, planning, management and strategy tools. The certified GSTRT professional has a broad knowledge of managing and building cybersecurity programs. An organization with a certified GSTRT employee can benefit from established credibility in meeting the requirements of board members, executives and the business. The GSTRT certification requires a candidate to complete one proctored exam with a passing threshold of 73%.

For more information, go to https://www.giac.org/certification/strategic-planning-policy-leadership-gstrt

GCPM

The GIAC Certified Project Manager (GCPM) is an advanced-level certification within the leadership and management domain. The GCPM certification ensures that a security professional has a strong knowledge of IT project management methodology, implementation and best practices. A certified GCPM professional is proficient with time and cost management, resource management, project risk management, procurement and project integration. The GCPM certification requires a candidate to complete one proctored exam with a passing threshold of 70%.

For more information, go to https://www.giac.org/certification/certified-project-manager-gcpm.

GLEG

The GIAC Law of Data Security & Investigations (GLEG) is an advanced-level certification within the leadership and management domain. It helps bridge the gap between the legal department and the cybersecurity team. A certified GLEG professional demonstrates the knowledge and skills specific to data security law, crime, policy, fraud, liability, contracts, active defense and IT security. A certified GLEG professional is well-informed about e-Discovery, data retention, compliance, business policies, third-party agreement, intellectual property and more. The GLEG certification requires a candidate to complete one proctored exam with a passing threshold of 70.7%.

For more information, go to https://www.giac.org/certification/law-data-security-investigations-gleg.

GSNA

The GIAC Systems and Network Auditor (GSNA) is an advanced-level certification in the management and leadership domain. A candidate looking to strengthen his or her risk analysis approach and technical audit techniques of information systems should apply for the GSNA certification. A certified GSNA professional has ample knowledge of network applications, auditing parameters in Windows and Unix environments, risk assessments and reporting. GSNA certification requires a candidate to complete one proctored exam with a passing threshold of 72%.

For more information, go to https://www.giac.org/certification/systems-network-auditor-gsna.

GIAC Certifications: Security Expert Category

GSE

The GIAC Security Expert (GSE) is an expert-level certification designed by cybersecurity experts to assess the technical and security expertise. The certification is framed to validate various skills required for an individual practitioner and security consultant. The GSE certification focuses on intrusion detection and incident handling. A certified GSE professional has multi-variate skills to direct an organization and strengthen its security system efficiently. A certified GSE professional works in a close group with top cybersecurity practitioners who help create a comprehensive understanding of complex technicalities and absorb viable information. The GSE certification targets professionals who want to enhance their technical approach and reach their career goals faster.

The GSE certification has some predefined prerequisites, which are available on the official GIAC website. A candidate must accomplish the certifications mentioned as a prerequisite to progress forward and attain the GSE certification. The GSE exam is a two-part exam-an entrance exam followed by a hands-on lab. The exam tests a candidate's skills in incident handling, intrusion detection and analysis.

For more information, go to https://www.giac.org/certification/security-expert-gse.

Other GIAC Certifications

GSSP-JAVA

The GIAC Secure Software Programmer Java (GSSP-JAVA) is a software certification. This certification validates a professional's ability to recognize security weaknesses in the current code and establish secure codes. A certified GSSP-JAVA professional has mastered the knowledge and skills required to manage common programming issues that can lead to major security issues. A certified GSSP-JAVA professional has also demonstrated skills to validate data, encrypt web applications, combat common web attacks, secure the Software Development Lifecycle (SDLC) and provide platform security. The GSSP-JAVA certification requires a candidate to complete one proctored exam with a passing threshold of 73.3%.

For more information, go to https://www.giac.org/certification/secure-software-programmer-java-gssp-java.

GSSP-.NET

The GIAC Secure Software Programmer .NET (GSSP-.NET) is an advanced-level programming certification. This certification validates a professional's expertise to write secure codes and determine any weaknesses within the software system. A certified GSSP-NET professional has demonstrated knowledge regarding authorization, encryption, data validation, .NET authentication, framework security, secure Software Development Lifecycle (SDLC) and session management. The GSSP-.NET certification requires a candidate to complete one proctored exam with a passing threshold of 66%.

For more information, go to https://www.giac.org/certification/secure-software-programmer-net-gssp-net.

VENDOR-SPECIFIC CERTIFICATION

At the beginning of this chapter, we identified two types of IT certifications: vendor-specific and vendor-neutral certifications. The certifications explored up to this point are vendor-neutral certifications from (ISC)[2], CompTIA, ISACA, EC-Council, Offensive Security and GIAC. Here we briefly explore vendor-specific certifications.

Vendor-specific certifications are considered highly beneficial for professionals seeking to meet an organization's needs to specialize in a particular vendor's solutions and tools; this is because vendor-specific certification focuses on an area of information security tied to a specific product.

The world of cybersecurity is evolving, and new technologies regularly emerge. Cybersecurity professionals are constantly juggling to acquire new techniques and tactics to improve an organization's security system. Vendor-specific certifications are a quick way to learn the practical implementation of new technologies.

Today, there are several vendor-specific certifications available. Our recommendation for aspiring cyber professionals is to define what a dedicated cybersecurity career journey looks like and then determine which certifications are relevant to achieve those goals. We explore how to map out a cybersecurity career journey in more depth in Chapter 7.

Chapter 5 Key takeaways

- There are two types of certifications available: vendor-neutral certifications such as the CISSP from (ISC)[2] and vendor-specific certifications such as Microsoft Certified: Azure Security Engineer Associate from Microsoft.

- For those starting a career in cybersecurity, obtaining a certification will assure employers that a professional has the knowledge base. Simultaneously, further training and work experience are required to develop the professional to become fully qualified in that certification.

- For those transitioning into cybersecurity mid-career, obtaining a certification is generally advantageous to build credibility within the field and progress faster toward defined career aspirations and goals.

Chapter 6: What are the Trends at the Forefront of Cybersecurity?

In this chapter, we look at some of the trends at the forefront of cybersecurity. These trends include technological advancements and the evolution of cybersecurity with these technologies and concepts. This chapter also aims to provide value-add information for those looking to specialize within a particular area.

The fact is we can no longer protect today's threats with yesterday's controls. In that light, businesses are rapidly evolving in response to the global environment. So much has changed just in the last year alone.

As we connect more and more devices from everywhere, the role of cybersecurity in protecting our rights, safety and privacy is urgent and prominent. Cybersecurity is no longer an option-it is evolving to become an intrinsic part of our day-to-day lives. It is a topic that businesses of all sizes care about and grave public concern as electricity grids, health devices, and vehicles become interconnected.

With the seemingly forced shift to digital, we are moving full throttle toward adopting new technologies. Technology leaders, such as CISOs, need to understand the inherent risk to their businesses quickly. In most cases, these risks are net new with previously unknown attack surfaces and vectors.

Business resilience to pandemics and other unforeseen circumstances

The boardroom conversation has changed since the emergence and peak of the COVID-19 pandemic. It now includes the urgency of business continuity, safety and resiliency in addition to profit generation in the face of economic adversity, brand image and operations. This change has happened because some organizations found themselves unprepared in identifying business impacts to profoundly unforeseen situations, such as the COVID-19 pandemic.

The following vital categories have now become clear mandates as we collectively settle into the new normal.

- **Secure environments:** With a remote workforce, it is imperative to provide employees a secure environment to do their work. Security leaders are looking beyond the traditional perimeter and endpoint security. A shift to cloud environments is no longer "nice to have" but is a vital business strategy.

- **Scalable and secure collaboration:** Teams have become widely distributed, and it is now a business necessity to have reliable, scalable and secure communication channels.

- **Changed security mindset:** Now more than ever, security awareness and training programs need to evolve to continually educate the workforce about the inherent security risks associated with working remotely rapidly.

- **Building foundationally resilient businesses**: It often takes a crisis for companies to swiftly approve a budget and mobilize resources to respond and return operations to business as usual. The COVID-19 pandemic is a stark example of a global crisis that exposed the fragility and consequence of lacking adequate business resiliency. Today, business leaders acknowledge the need to build a resilient business; however, many still struggle with understanding how to do just that. Companies need to embed resiliency at the foundational level of their operations. Where applicable, the business should establish a business continuity program focused on the building blocks of business continuity, emergency response, disaster recovery and crisis management aligned with industry frameworks and standards such as ISO 22301.

This chapter explores several emerging trends specific to technological advancements, business benefits and where cybersecurity plays a critical role. As future cybersecurity practitioners, consider how these trends may impact the specialization you choose to endeavor. In no particular order, we explore trends at the forefront of:

1. Leap to the Cloud
2. Artificial Intelligence and Machine Learning
3. Internet of Things (IoT)
4. Operational Technology (OT)
5. Robotic Process Automation (RPA)
6. Robotics
7. Drones
8. Connected and Autonomous Vehicles
9. 5G
10. Blockchain in Cybersecurity
11. 3D Printing
12. Data Protection
13. DevSecOps
14. Threat Vulnerability Management (TVM)
15. Software-Defined (SD) Security
16. Security Orchestration Automation and Response

LEAP TO THE CLOUD

The need of the hour is cloud security. The COVID-19 pandemic forced businesses globally to shift to a fully remote workforce and change IT operating models almost overnight. In turn, security teams experienced dramatic and wide-scale changes to their environments, including accelerating cloud adoption to enable operational continuity and resilience. The shift to the cloud is no longer an option-it is a business mandate.

Cloud security is about addressing the confidentiality, integrity and availability of data stored on systems hosted outside the primary data centers of an organization, most likely by a third-party cloud service provider. These cloud service providers host organizational data on highly elastic and scalable infrastructure, also known as the "cloud."

To that end, a comprehensive program and strategy to embed security throughout the enterprise cloud life cycle starts at the contract level, to data protection controls, to business continuity and beyond. Integration and interoperability of platforms and applications bring about another important question around the secure transition to the cloud. The strategy increasingly needs to include a structured approach to embedding audit, legal, compliance and security requirements; this requires an integrated governance mechanism.

The focus on the cloud is ubiquitous. However, the readiness of some businesses to shift to the cloud is not, and there are inherent gaps toward a secure transition. Gaps in policies, user training, access management are almost always found within organizations preparing to make the cloud leap. Cloud Access Security Brokers (CASBs) can address these gaps.

CASBs have the capabilities to mediate the existing control gaps by providing a comprehensive view of data, devices and users by integrating technologies that provide operational visibility into data flow and user behaviors. Some of the native capabilities offered by cloud service providers include governance and visibility, logging and monitoring by default, container security and isolation, identity and access management strategies (e.g., Multi-Factor Authentication [MFA], Single Sign-On [SSO] and Privileged Access Management [PAM]) and data life cycle security.

The cloud provides elastic and highly flexible business environments in these times (which are pretty uncertain, to say the least). The ability to deploy enterprise information resources securely and at scale with reduced cost over time and increased visibility certainly explains why organizations from all industries and sizes with varying regulatory mandates embrace the journey to the cloud. Overall, the organic move to the cloud requires a shift in traditional security viewpoints—to change to a more governance-based model and a change in the necessary skills—requiring a move from perimeter-based to more data-centric-based security controls.

Cloud security concerns vary in both privacy and security. Some of these concerns include:

- **Data protection:** From a technical and administrative perspective, proper protection of critical and sensitive data on the cloud is a concern. Technical controls would include logical access and access based on "need to know" and "least privilege." Administrative controls consist of the soft controls established concerning protecting the data and the privacy considerations embedded to consider the data's physical/geographic location.

- **Misconfiguration:** There are security concerns around misconfigurations, such as misconfiguration of storage mechanisms, lack of proper credential management and lack of access management. These misconfigurations can assist a threat actor's ability to gain unauthorized access to an organization's data.

- **Cloud security monitoring, threat detection and predictive analytics:** The presence of ineffective monitoring and intrusion detection strategies can significantly harm cloud operations. Security experts need to predict any potential vulnerabilities to avoid uncertainties comprehensively.

There was a heightened acceleration to shift to the cloud following the crisis of the COVID-19 pandemic. Cloud security, in turn, has become a significant part of an organization's overall security program. Next, we look at the advantages of adopting the cloud and where cloud security is headed.

What are the advantages of adopting the cloud?

- **Cost-effectiveness:** Short-term and long-term cost is a primary consideration when businesses adopt new technology. The budget can make a massive difference in cybersecurity and investment in appropriate controls. The main benefit of cloud technology and cloud security implementation is cost-efficiency. Services on the cloud are usually pay-as-you-go, which does not require a lump sum investment (i.e. Capital Expenditure [CAPEX]). Organizations can also adopt investments in the cloud within their Operational Expenditures (OPEX).

- **Operational flexibility:** Adopting various technologies can burden an organization's IT department resulting from increased resource time and effort to avoid potential risks. Acquiring assistance from a third-party service provider to maintain an organization's operations (including security) can lessen potential operational burdens. The practical implementation of cloud security provides flexibility to an organization to focus on other crucial functions by cutting down in-house resource time and enable proactive monitoring through trusted third-party cloud service providers.

- **Security at the core:** A primary task of a cloud server provider, or CSP, is providing operational resilience by enabling secure infrastructure. Resiliency forms the core of the CSP offering and is inherently more efficient than a conventional on-premise system where an organization needs to carve out time between securing the infrastructure and other IT concerns. RapidScale claims that 94% of businesses saw an improvement in security after switching to the cloud, and 91% said the cloud makes it easier to meet government compliance requirements.[35]

- **Business continuity:** Today's digital businesses have minimal tolerance to disruptions, and any downtime in services means lost revenue and negative reputational impact. Strategic recovery priorities based on a collaborative business impact analysis is again at the core of cloud capabilities enabling faster recovery from unforeseen situations

Where is cloud security headed?

As cloud adoption continues to grow and return on investments are realized, advancements in technological innovation and automation will continue to happen in the cloud. While the cloud can be more secure than a traditional on-premise infrastructure, the control and ownership model can become complicated. Modifications to governance models, security operating models and playbooks are needed to support the complexities of the shared ownership model, including the challenges with cloud forensics.

The lack of or ever-changing compliance and regulatory mandates can also become complicated, requiring cloud consumers to plan and orchestrate how to meet compliance with due diligence and due care effectively. Also, as attack vectors move to the cloud, the threat surface inherently increases and misconfigurations may mean havoc. Enforcing a detailed look at cloud configurations for proper mapping to specific applications and data requirements is needed.

Artificial Intelligence and Machine Learning

Artificial Intelligence (AI) is a broad area of computer science that seeks to combine human-like cognition with the agility of automation. Machine Learning (ML) is a subset of AI which, through advanced algorithms that improve automatically with experience, continues to bring advancements to today's digital age. AI also brings unlimited possibilities for augmenting human discretion with intelligent decision-making based on contextual correlation by connecting data points. Unfortunately, the adversary also leverages these advancements to launch more sophisticated cyberattacks. In turn, this creates the need for cybersecurity professionals to understand relevant use cases to defend their organizations effectively.

The progress of AI and ML continues to benefit cybersecurity practitioners with improved control mechanisms to defend against malicious activities. By removing human involvement, AI and ML applications enable the automation of threat detection and response capabilities. Big data analytics can gain insights from vast data sources, such as threat intelligence and system logs, which provides better prevention and recovery strategies.

In cybersecurity, ML helps in several areas, such as prioritizing and identifying risk, conducting analysis to reverse engineer malware quicker than ever before, and acting as a guiding body for incident response teams to detect overall intrusions. The use of coding for normalization of source data combined with specific domain expertise is increasingly

ML is becoming a crucial part of an organization's security capabilities. For example, ML helps identify zero-day vulnerabilities while also protecting infrastructures by looking at patterns and behaviors rather than signatures. AI/ML implementations also help with malware identification and thwarting spreads by determining a malware's risky behavior, such as when someone clicks on a phishing email.

What are the advantages of using AI and ML within cybersecurity?

- **Big data processing and deep learning:** Organizations that manage large volumes of data can use cutting-edge technologies to gain insights from datasets. ML technology can sort through these large volumes of unrelated datasets and spot the anomalous ones to identify and alert on threats and, where possible, take remedial action before those threats inflict mayhem. An organization with the advanced implementation of ML technology can combat, prevent and predict vulnerabilities effectively.

- **Threat intelligence and insights:** ML technology boosts an organization's ability to use patterns and behaviors to gain insights on Indicators of Compromise (IoCs) and anomalous activities. The segregation of normal activity adds to the ability of ML to detect abnormal ones. ML technology contributes to determining legitimate vulnerabilities at the initial stage before spreading further. Through ML implementation, alerting mechanisms can remember previously encountered vulnerabilities and restrict threats from surfacing again.

- **Reduction of mundane tasks:** The application of AI and ML is increasingly assisting in automating manual and operationally heavy security tasks and product integrations. In turn, this leads to improved threat detection and remediation activities. For example, anyone that has worked in the threat and vulnerability management (TVM) space recognizes the highly manual processes involved. Each activity involves substantial human intervention from scanning, risk rating, testing and patching. TVM automation is one area where AI and ML are helping immensely. Another area is the highly manual Security Operations Center (SOC) Level 1 activity that generally involves triaging security alerts and assigning those alerts to the appropriate parties for remediation.

- **Strategic investment:** Organizations actively look for long-lasting and cost-effective solutions to enhance and maintain their cybersecurity posture, providing tangible ROI with minimal human intervention. AI/ML is an adaptive technology that requires minimal human mediation. The technology has the potential to update itself, requiring limited manual upgrades. For example, a SIEM solution can leverage AI/ML implementations to collect actionable security intelligence and combine that intel with security incident management playbooks to improve alerting and monitoring.

Where are AI and ML technologies headed?

Implementing AI and ML technologies continue to create a convenient, connected and safe world for us. However, threat actors also use these technologies to automate their efforts. According to a BuiltIn.com article called *Machine Learning Cybersecurity: How It Works and Companies to Know* by Gordon Gottsegen, CTO of AXA IT North Europe Yorck Reuber told Darktrace: "We're not being attacked by human beings anymore. Computers are attacking us; software is attacking us. The only way forward is using artificial intelligence."

Although AI/ML aids in solving the data-centric threat surface, human cognition and domain expertise are needed. Over-reliance on AI and ML can create a false sense of safety, which is why in addition to appropriately applied algorithms, cybersecurity experts are required.

Internet of Things (IoT)

Internet of Things (IoT) is changing the way we live and work. Sometimes referred to as connected devices, IoT are devices with sensors, a network connection and can transmit, store and analyze data. Examples in our home and office include intelligent appliances, smoke detectors and door locks. Beyond our homes, IoT is revolutionizing the manufacturing, automotive, transportation, health care, energy and agriculture sectors.

Any device consumers purchase that connects to the Internet can be vulnerable to cyberattacks. Smartphones, smart TVs, home assistants (e.g., Amazon Alexa, Google Assistant), security cameras and baby monitors can have exploitable vulnerabilities if not configured and patched correctly.

Product developers are making IoT devices enticing to both businesses and consumers, and these are rapidly making their way into our everyday lives. For product developers, there is a rush to bring these devices to market to start making money. However, security and privacy measures are often an afterthought and not baked into the code or engineering building blocks that run these devices.

There are many serious risks should IoT devices become comprised, and control of such IoT devices fall into the wrong hands. Among the most troubling examples are the smart health devices that help keep people alive by providing continuous monitoring of vital statistics. Hackers have the potential to compromise devices like pacemakers and medical infusion pumps. Exploitable vulnerabilities are present in these devices, resulting in remotely administering lethal doses, stopping devices through denial of service or theft of sensitive information.

Cybersecurity professionals who establish specialization in IoT security will have an exciting and meaningful career path by contributing to the enhanced safety in the development and protection of these devices.

What are the advantages of IoT technology?

- **Improved and informed decision making:** Organizations actively look for new revenue streams and business opportunities to be more profitable and future-ready. Having more interconnected devices can afford better decision-making. For example, a retailer may know what products to buy and when based on predictive customer demands. IoT devices tend to accumulate data from various networks and utilize advanced analytics techniques to derive rational growth opportunities. An organization's customer strategy department can use IoT devices to systematically sample extensive customer demographic data to plan an upcoming campaign. The growing need and significantly positive impact of IoT devices on a business make it one of the cornerstones of the overall digital evolution. However, synergizing this growth with security and privacy solutions is required to restrict IoT's negative impact.

- **Attain greater return:** A robust IoT network is crucial for organizations as they observe the market's growing demand and act rapidly to enhance customer satisfaction. Data collection and analysis can determine past trends and forecast upcoming scenarios so that organizations can make informed decisions and act upon them faster. Most organizations use IoT devices to enhance their internal operations and keep a check on their liabilities. IoT technology uplifts an organization's operational abilities through its connectivity and ultimately helps in attaining greater returns.

- **Improve monitoring:** IoT devices have the remarkable ability to connect with various servers within a network, which can help monitor an organization's operational activity. Monitoring has many layers, and organizations have different techniques to do it. IoT devices can help track employee activity, providing user behavior analytics and potentially predicting security breaches. IoT devices can also support the maintenance of physical security for an organization.

- **Enhanced automation and control:** As more physical devices become centrally controlled with machine-to-machine interaction, automation and orchestration are significant advantages in minimizing mundane tasks, thus enabling resources to focus on more relevant activities while also saving time and overall cost.

Where is IoT headed?

With increasing use cases emerging constantly, more and more organizations feel confident in investing and adopting IoT strategies to enhance business profitability. However, the risks associated with IoT cannot be ignored. Businesses need cybersecurity professionals to provide their expertise to safeguard the emerging technology and ensure it is safe to use in businesses and homes.

It is somewhat eye-opening that most articles we come across while researching IoT indicate privacy and security as disadvantages. However, it does not have to be this way. By including security design principles and privacy-by-design concepts throughout the development life cycle, these devices can continue to improve the way we live and work safely and while protecting our privacy.

OPERATIONAL TECHNOLOGY (OT)

Operational Technology (OT) is specialized software or hardware that manages industrial equipment, assets, processes and events. OT differs from Information Technology (IT) where OT is deployed in line with a direct business function of an organization.

Examples of OT include Industrial Control Systems (ICS) and Supervisory Control and Data Acquisition (SCADA). The electricity that reaches our homes moves from generation stations to transmission stations to distribution stations, all of which have OT components. Similarly, organizations within the defense, manufacturing, energy and resources use OT for most of their core business functions. At the heart of OT are Programmable Logic Controllers (PLCs) that provide the required functionality.

Like IT, OT components and the IT portion of the network that manages the OT components can be misconfigured, hacked and potentially compromise people's safety and security. The Tampa Bay Times reported on February 8, 2021 that there was a hack at the Oldsmar water treatment facility, which is located in the U.S.-based Florida Water Plant. Based on the article, an attacker used remote access to the system to change the level of sodium hydroxide, more commonly known as lye, in the water from 100 parts per million to 11,100 parts per million, trying to "poison the water supply."

According to various sources, the compromised servers were using an outdated Windows 7 operating system. It is important to note that Microsoft already ended support for the Windows 7 operating system back in early 2020, which means that no new security patches are published by Microsoft, leaving these systems exposed to new cybersecurity vulnerabilities. This example is eerie as it raises the concerns the convergence of IT and OT brings. Although the intersection means a more connected and centrally managed OT environment, it presents new security challenges.

As OT organizations increasingly adopt IT technologies to become more agile and competitive, more ICS, SCADA and intelligent sensors connect to IT data centers. Inherently, OT and IT professional skill sets evolved separately and only recently has there been the need to bring both professional skills together. The demand for cybersecurity professionals who specialize in OT security continues to rise with no shortage of opportunities to thrive.

What is the need to enable comprehensive cybersecurity measures within OT networks?

- **Disruptions and safety concerns:** Cyberattacks on ICS and SCADA systems that run our power grids and public transit systems can potentially impact the safety of the workers and our communities. Due to limited interaction with IT systems, OT systems lack visibility, analytics and timely vulnerability management. In turn, this creates a significant threat surface requiring the development of a strategic and uniform cybersecurity strategy.

- **Organization-wide visibility:** Cybersecurity technologies and operations that can provide enterprise-wide visibility are critical to proactive monitoring and determining security threats. A holistic overview of organizational function on a single platform makes the monitoring process a lot easier. The blind spots within an organization's system can limit overall growth while increasing financial, regulatory, operational and reputational stress.

- **Building an ecosystem of trust:** The collaboration and integration of IT and OT solutions can help gain situational awareness and secure operational solutions for seamless growth. OT helps leverage current organizational investments and integrate IT security technologies, such as next-generation firewalls, IT service management and Security Information and Event Management (SIEM) systems.

Where is OT headed?

The fourth industrial revolution, or commonly known as Industry 4.0, centers around interconnectivity, automation, machine learning, unifying operational data streams and resilient decision-making. Industrial IoT (IIoT) and smart digital technology are transforming OT and revolutionizing the way industrial businesses operate and grow; this includes optimization in asset tracking and supply chain management, predictive maintenance and diagnostic analytics and more.

With this evolution, the challenge businesses face is filling vacancies for the net new set of necessary professional skill sets that converge IT and OT expertise. Characteristically, IT teams are not trained on operational technology and OT teams are not trained on IT security strategies. Professionals whose skill sets and aspirations can blend between the two are needed now and into the future of Industry 4.0.

ROBOTIC PROCESS AUTOMATION (RPA)

Robotics Process Automation (RPA) is a technology that alleviates manual business procedures by integrating automated processes. The integration of an automatic process adds to an organization's efficiency in executing operations faster and potentially with more minor errors. RPA allows configuring computing software to integrate and emulate human actions by interacting with digital systems to achieve organizational processes seamlessly. The augmentation of RPA technology into an organization's system helps capture data at an impeccable speed by establishing and propagating error-free patterns.

RPA can also manipulate applications and imitate human actions in an automated way. RPA triggers responses, mimics human actions, interprets and captures data, and communicates with other systems within a network to produce a better output. Various employees have to perform monotonous tasks, limiting their efficiency and the courage to perform better. To make better use of the workforce, organizations adopt RPA to execute repetitive tasks. The automated system, often termed a Robot in an RPA system, is active and performs a job with minimal errors.

RPA is a powerful mechanism that enhances an organization's ability to standardize and streamline various process-oriented business activities. RPA technology is gaining rapid popularity as it is considered a core competence within an organization's digital transformation journey. RPA is one of the newest technologies that are still subject to a variety of testing processes. As with many other advancements, RPA will scale as it becomes more turnkey for business advancements with increased adoption. Understandably, many organizations are captivated by the RPA technology and are closely observing its long-term benefits. For now, these organizations have included RPA technology on their technology adoption roadmap.

What are the advantages of using RPA in cybersecurity?

- **Fast incident response:** Humans are one of the crucial elements in inducing risk within an organization. The risk exposure caused can be unintentional or based on ignorance. The adoption of RPA has immense potential to minimize the burden of performing repetitive and error-prone tasks, allowing a workforce to focus on more essential tasks. The automated vulnerability detection and mitigation capabilities of RPA strengthens an organization's security posture. Through RPA, organizations can automatically roll out patches and updates to systems to address vulnerabilities. Another example where RPA has gained popularity is automating the security operations center's Level 1 and 2 tasks, which revolve primarily around incident triage, categorization and assignment.

- **Optimizing SOAR capabilities:** Security Orchestration Automation and Response (SOAR) is a powerful automated security technology that plays a crucial role in automating security incident detection and response activities. SOAR technologies significantly optimize the operational capabilities of an organization by freeing humans from significant low-level responsibilities. RPA is at the core of SOAR technology, as it is combined with automated solutions to add to an organization's ability to respond to security incidents. The faster evaluation of security issues plays a significant role in an improved security posture.

- **Threat intelligence:** RPA is an essential part of threat intelligence as it uses automated solutions and technology to determine potential indicators of compromise (IOCs). The multivariate approach of mitigating risk through dynamic threat analysis is substantial in providing a competitive edge to organizations remaining on top of security threats.

- **Alleviated security posture:** Automation expedites the process of vulnerability detection and system patching to avoid severe damage. It can take a significant amount of time to determine a viable action plan through manual procedures. RPA contributes to the faster determination of threat action plans and seamless implementation of solutions to avoid vulnerabilities. More coordinated attacks can shake the security infrastructure of an organization and cause a significant financial setback. RPA technologies provision for innovative operational approaches to optimize outcomes.

Where is RPA technology headed?

Brandessence market research released a report in January of this year called *Robotic Process Automation Market 2021,* stating the RPA market size is expected to see significant growth reaching up to $18339.95 million USD by 2027. While RPA is still in its development stage, organizations globally will continue to embed RPA technologies at the desired scale to automate their business output by up to 45%.[36] RPA technology is beneficial as it adds value to an organization's security system by automatically gathering and analyzing data while relieving the workforce from executing mundane tasks. Cybersecurity products will also adopt RPA, and the need for cyber professionals within this space will continue to grow.

ROBOTICS

The deployment of robots, especially in manufacturing industries, is becoming an increasingly common practice. Mundane procedures and repetitive actions can easily take a toll on employees and affect productivity. Robots can emulate human activity efficiently, optimizing results and providing organizations with leverage to use their workforce for more valuable tasks such as supervision, administration and strategy development.

With the growth of robotics systems, an increasing number of security flaws have come to light, including physical safety, software development vulnerabilities and data protection. Organizations that are considering adopting this technology are aware of the economic viability, including better hardware, improved communications and advanced algorithms. However, organizations are also reserving implementation of robot solutions due to cultural acceptance of robots and the need for more use cases, a better understanding of ROI and improvements in security risks.

What are the advantages of using robotics and automation technology?

- **Productivity**: Every organization has specific expectations regarding productivity, often not met due to increasing workload or human errors. A well-programmed robot is aligned to perform operations seamlessly with minimal (if any) errors. The execution of tedious processes with utmost precision leads to excellent efficiency. Organizations then gain enough leverage to smartly re-allocate resources to more valuable tasks and, as a result, potentially gain a competitive edge.

- **Speed and agility:** A robot is an automated device that works without feeling stressed or burdened in complex circumstances. The high pressure to meet deadlines is lifted off employees' shoulders. Robots can work day and night, which is an incredible attribute for an organization to complete mundane tasks faster. However, as software code underpins robotic efficiency, it is inherently prone to security flaws in the code, which can be exploited.

- **Safety:** Some tasks are characteristically unsafe for human beings. The earliest instance of the usage of robots was to complete highly hazardous tasks, such as diffusing a bomb. More and more, robots are used to save human beings from dangerous exposures.

Where is robotics technology headed?

The COVID-19 pandemic forced some companies to choose between 1) finding ways for employees to do their jobs safely; 2) shut down entirely; or 3) displacing workers with automated robot technology. "Robophobia" is once again on the rise, understandably so.

Not only did the pandemic force digitalization almost overnight, robotic automation is rapidly becoming a viable option for organizations, whether we are ready for it or not. As robots continue to transform our future, the need for cybersecurity expertise and solutions will also continue to rise.

DRONES

The use of drone technology has ballooned over the past few years-from express deliveries, aerial photography, search and rescue, surveillance and disaster management to name a few. The latest headline about drones, also known as unmanned aerial vehicles (UAVs), suggests that UAVs may be used to deliver coronavirus medical supplies to clinics at difficult-to-reach locations, including rural communities, to help reduce exposure and free up hospital beds.[37]

While drones have historically been associated with military forces, the transport, agriculture, logistics, and commercial sectors are fast adopting drone technology. The next generation of UAVs is smart drones designed with artificial intelligence and computer vision technology readily capable of facilitating autonomous operations, self-monitoring and accurate sensors while meeting aviation compliance requirements.

Because UAVs are free-range Internet of Things (IoT) devices equipped with wireless communications, these are susceptible to cybersecurity and privacy threats, such as hijacking to steal sensitive data and supply chain threats where flight data may be sent back to manufacturers abroad. Even hobby drones using Raspberry Pi computers can sniff Wi-Fi signals and listen into communications at remote critical infrastructure establishments, such as power stations and offshore oil rigs.[38]

While the adoption of drone technology at a mass scale is still in its infancy, it essential for cybersecurity practitioners to understand the benefits, threats and future of this progressively innovative technology.

What are the advantages of using drone technology?

- **Remote monitoring:** Drone technology is vital in providing a cost-effective and efficient bird's eye view of large areas or difficult-to-reach facilities, such as (and not limited to) port and ship facilities, off-shore oil rigs and wildlife monitoring. The business benefits of aerial surveillance and remote monitoring include real-time monitoring of operations sites and enabling staff to recognize and respond to issues quickly, which in turn improves health, safety and security measures.

- **Perimeter security:** Drones provide significant benefits beyond traditional physical security devices. For example, mounted cameras for video surveillance are fixed, whereas a drone is mobile and provides a 360 view. Security guards or human patrols are kept out of harm's way with remote aerial surveillance and stealth mode (near silent) capabilities of drone technology. And detection software enables real-time recognition and reporting of perimeter breaches. The effectiveness of perimeter security using drones is further enhanced with adequate aerodynamic knowledge, human operator training, and monitoring and response processes in place.

- **Low-carbon footprint:** With the sharp rise of drone technology for express commercial delivery of products, studies have shown drone technology has the potential to reduce energy use and greenhouse gas emissions compared to diesel-powered land, sea and air transportation. Advances in drone technology has also made it possible for drones to travel farther at affordable prices and easily accessible for consumers and businesses alike. These benefits will continue to propel the demand for drone technology. Cybersecurity practitioners should be increasingly aware of how to effectively maintain the safety, physical security and data protection of drones.

Where is drone technology headed?

The COVID-19 pandemic has catapulted the demand for drones and aerial fleets in the transport, agriculture, logistics, healthcare and commercial sectors globally. While advancements in drone technology are revolutionizing the supply chain, it has also led to the rise in IoT device security issues, including physical security, human safety and privacy protection.

The probability and frequency of the malicious use of these drones are high, with potentially devastating impact in the military, civilian and terrorism domains. Furthermore, drones can malfunction mid-flight and cause injuries or fatalities. Therefore, organizations must practice due diligence and due care when using or considering drone technology to implement detective, protective and preventative security countermeasures.

These countermeasures can include regularly conducting a drone vulnerability and risk assessment; implementing physical, network communication and data protection security controls; and planning and periodically testing a drone incident response plan. Cybersecurity practitioners must think like a hacker when it comes to safeguarding organizations that use this progressive technology. Thinking like a hacker includes understanding how the adversary could intercept communications or hijack a drone system to steal critical data, recording drone system vulnerabilities and implementing a drone security program.

CONNECTED AND AUTONOMOUS VEHICLES

Connected and Autonomous Vehicles (CAVs) are driverless vehicles that use automated driving systems to sense their environment and travel to programmed destinations with little to no human involvement. Examples of driverless vehicles include autonomous haul trucks at mining sites, passenger cars and more. Today, CAVs are still in its research and development stage worldwide.

For autonomous haul trucks deployed at mining sites, a control system is placed on the trucks, and a computer control room controls each truck's destination, speed and operation. These autonomous haul trucks are connected to other equipment in the mine, creating an autonomous hauling system whereby each equipment is housed with a Global Positioning System (GPS) that way trucks are aware of where each equipment is relative to each other. An autonomous haul truck travels along predefined courses to assigned destinations at the mine site.

Similarly, an autonomous passenger car has various sensors throughout, which helps establish and maintain a comprehensive map of its surroundings. The radar sensor evaluates the distance of the car from other vehicles on the road. A video camera within the car's internal system also evaluates the traffic and alignment on the road.

An essential component of autonomous vehicles is the Light Detection and Ranging (LiDAR) sensors, which work intelligently to bounce off the light from the road and other vehicles to measure the distance and situate the vehicle in a safe space from other vehicles. Ultrasonic sensors are located in the wheels of the vehicle, which analyze the surrounding curb when parking. Autonomous cars have sophisticated software functions to process destination plots, send instructions and evaluate the sensory input. The car's software manages the actuators, which controls braking, steering and movement.

The development of autonomous vehicles seems attractive to various transportation organizations deeply involved in integrating this technology within their vehicles. However, it is difficult to ignore the cybersecurity implications that are a part of CAVs. The use of sensory controls within an autonomous vehicle is alarming for many manufacturers. There are high chances of threat actors breaking into the vehicle's computer software and controlling its functions. Several operational technology and software devices within autonomous vehicles can carry manufacturing defaults and vulnerabilities that threat actors can exploit. A protected and well-managed security environment can enhance the efficiency of an autonomous vehicle and make it more viable for public use.

What are the potential benefits of fully autonomous vehicles?

- **Road safety enhancements:** Some leading causes of vehicle accidents are impaired driving, speeding, distraction and unbelted vehicle occupants. The IoT technology used within autonomous vehicles enhances vehicle connectivity and sensory with other vehicles on the road. The sensory controls in CAVs automatically calculate the distance from other vehicles and obstacles relative to the position of the driverless vehicle. The automated connectivity technology within driverless vehicles help reduce accidents by helping clear the human error dependency.

- **Reduced commute time, reduced congestion and increased productivity:** The sensory mechanism and use of intelligent data in CAVs enable faster reaction and improved driving performance compared to humans. In turn, autonomous cars can help save time spent on commuting and parking, thus providing more time for productivity. The road infrastructure and CAV connectivity map to the GPS helps the driverless vehicle evaluate the traffic concentration within specific areas and take the vehicle to low traffic areas. Furthermore, autonomous cars use internal software and sensory technology to automatically park vehicles close together freeing up space.

- **Subsequent cost savings:** While driverless cars are not yet available to purchase at an affordable market price compared to conventional vehicles, there are subsequent cost savings. These savings may include car insurance, reduced or no fines for driving or parking violations, automated delivery and ride-sharing. Because driverless vehicles help protect the driver, there may be additional cost savings in healthcare and vehicle repairs from accidents.

Where is connected and autonomous vehicle (CAV) technology headed?

Connected and autonomous vehicle (CAV) technology is still in its research and development phase. Its growth in consumer demand is dependent mainly on regulation, sensor hardware costs and public perceptions of safety. Based on research data available today, wide-scale adoption of driverless vehicles may include benefits such as increased driving performance, fewer accidents, reduced congestion, subsequent cost savings and improved time savings.

As driverless vehicle technology matures and advances over the next several years, cybersecurity must be embedded in the engineering and development process. Cybersecurity practitioners must understand how this technology works today, know the various levels of autonomy defined by the industry, understand exploitable vulnerabilities and recommend security strategies to safeguard its interconnected technology from cyber threats. Vulnerabilities may include insecure network connections, difficult-to-find vulnerabilities in legacy software and open-source code, spoofing vulnerability of vehicle sensors and malicious code embedded in the CAV mapping system.

During this research and development stage of CAV technology, policy makers and lawmakers are playing catch up to ensure proper regulations are implemented before the broad deployment of fully automated driverless vehicles on autonomous transport networks across smart cities.

5G

5G is the fifth-generation technology standard for cellular networks, which enhances mobile broadband access. 5G works by segregating geographical locations into small areas. Each division is called a cell. The cells are interconnected to provision for vast and fast connectivity. 5G is designed for consumption in larger regions and developed for efficient means of broadband access.

5G has the potential to meet the increasing demands of consumers for faster connectivity to the Internet. The development of technologies, such as the Internet of Things (IoT) and the Internet of Medical Things (IoMT), is expected to progress by using 5G technology. Many organizations are upgrading their infrastructure to fit 5G technology with their existing networks and IoT devices. The build of smart cities is also possible through the use of fast connectivity networks.

5G technology helps facilitate 24/7 connectivity enabling fully functional and operational smart homes, smart factories and smart hospitals. All the while, 5G networks have introduced an expansive cyberattack surface through the need for new network architectures, software virtualization, expanded bandwidth and IoT proliferation. Exploitable vulnerabilities of 5G include malicious software introduced within the supply chain, misconfigured or improper deployment creating vulnerable networks, and legacy vulnerabilities from generations of wireless networks.

What are the benefits of 5G technology?

- **Fast connectivity:** The world is evolving with new technology, and innovations replace the old ones at an impeccable speed. Organizations have to keep up with the technology evolution to enhance their business efficiency. 5G is one of the fastest connectivity technologies available today.

- **IoT:** Fast connectivity and vast accessibility are focal points in enhancing the efficiency of IoT technology. 5G has built a strong foundation for more devices to enter a network and generate a significant amount of data.

- **Speed in remote work:** With remote work as the new normal, businesses continue to observe the impact on workforce productivity and output. Early data shows productivity has increased for some businesses with work-from-home employee arrangements. The speed of 5G connectivity offers employees the same performance as office technology.

- **Low latency:** Latency, synonymous with delays from seeing results on devices, is the amount of time it takes for a signal to reach its destination and back. Low latency means short-to-no delays, which is a game-changer for 5G technology. For example, the shift to an online culture with 5G and low latency is empowering physicians to attend to their patients remotely through innovations in telemedicine.

Where is 5G technology headed?

5G is more than just faster connections for smartphones. 5G connects billions of IoT devices to the Internet faster, thus allowing simultaneous connections between many IoT systems, such as connected and automated vehicles (CAVs) and wearable technology in healthcare. 5G technology is revolutionizing the healthcare industry whereby physicians can monitor patients remotely through wearable technology such as fitness trackers, smart health watches, wearable ECG, blood pressure monitors and biosensors. Wearable electronic devices collect sensitive personal health information, and the potential for compromise of medical data or identity presents a real threat.

Beyond healthcare, 5G will continue to play a critical role in developing smart cities, autonomous vehicles, telemedicine, connected factories, product development, education, urban infrastructure, AI-enabled customer service and more. With the future of 5G taking shape right before our eyes, the market indicates a strong demand for cybersecurity practitioners to help protect organizations adopting this technology now and into the future.

BLOCKCHAIN IN CYBERSECURITY

In response to data protection, blockchain technology has attracted many organizations due to its promising ability to secure data while keeping the integrity of the data intact in a highly distributed environment. Simply defined, blockchain is a secure way to create tamper-proof logs or transaction records of activities.

Some examples of where blockchain technology is used include:

- safekeeping of electronic medical records in the healthcare sector;
- automating claims and verification of coverage between companies in the insurance industry;
- exchange of digital currency in the financial sector; and
- digital identity in law enforcement agencies such as international border services

In cybersecurity, the blockchain concept is incorporated to restrict fraudulent activities via a consensus mechanism. The consensus mechanism is built based on a peer-to-peer network, where transactions are chronologically recorded and grouped into blocks that are cryptographically secured and organized in chains. The consensus mechanism consists integration of a distributed technology ledger (database) that detects data tampering by way of cryptographic hashing. Any change made to the database is tracked as a new transaction with different time stamps. As such, operational resilience, transparency, immutability and audibility are built-in. A distributed database is a blockchain mechanism that is applied in a public or private domain. This concept differs from a centralized structure in which data is gathered in the form of large databases.

Blockchain technology allows transactions, such as exchanging sensitive data between trusted parties, to be conducted in a non-altered methodology. Distributed public ledgers are based upon the blockchain strategy containing immutable data in an encrypted and secure way; this works in favor of organizations to secure their data correctly and closely monitor all the potential vulnerable activities. The Distributed Ledger Technology (DLT) is gradually making its way to the list of technologies organizations seek to integrate after blockchain's use in cryptocurrency and bitcoins.

What are the advantages of using blockchain technology in cybersecurity?

- **Traceability:** When a transaction is made, it is time-stamped and digitally signed. This procedure enhances an organization's ability to trace a particular transaction to a third party at a specific time in the public domain. The traceability feature of blockchain is highly beneficial to organizations as an unauthorized activity can be tracked when it disrupts the chain. A blockchain's functionality ensures the reliability of a security system as each transaction is cryptographically associated with the user. Blockchain technology has impeccable audit benefits as it provides a high level of transparency and immutability.

- **Passwordless authentication:** Blockchain technology can allow an organization to authenticate users and systems without passwords based on decentralized identity and access management. One way this can be made possible is by enabling a trust mechanism that uses chains created from user biometrics embedded into a secure mobile application. As this minimizes human intervention during the authentication process, the attack surface is significantly reduced. The conventional system's primary weakness is the simple login mechanism and centralized security architecture. The security efforts of an organization are futile if they use system passwords that are easy to crack. Blockchain technology empowers an organization's security efforts by offering strong authentication and reducing the single point of failure.

- **Decentralized storage:** Users of blockchain can maintain and store data within their systems in a peer-to-peer setting; this acts as an assurance that the blockchain will not collapse in case a single node fails. If a threat actor attempts to interfere with the blockchain and its mechanism, each system in the blockchain analyzes the data and determines systems that differ from the other systems. If any system differs within the blockchain, the chain simply excludes that system from the chain. The distinctive design of the blockchain does not recognize any system as a centralized authority. Every user on the blockchain can continually verify the data while ensuring there is no case of data tampering or alteration.

Where is blockchain technology headed?

Blockchain technology may soon replace traditional digital records. In industries like finance and healthcare, blockchain technology is advancing data immutability and transaction transparency requirements to protect from fraudulent activities, data loss properly, and customer privacy are at the forefront. As global supply chains become highly complex, multinational giants like Walmart use blockchain technology for supplier transparency across the entire food chain. Another example is in the real estate industry, where smart contracts are increasingly used for the fast and secure execution of high-value agreements between multiple parties.

Because of its transformative value-add, including private data transmission, open-source, redundancy, anonymity security and decentralization, experts see explosive growth adoption of blockchain technology for years to come. As blockchain technology adoption grows, security questions about reliance on private keys, adaptability, and quantum computing challenges are rising. Therefore, cybersecurity professionals need to keep up with this innovation to help answer these pertinent questions.

3D Printing

On-demand 3D printing has taken the world by storm. According to a Deloitte Insights article, 3D printing is expanding at a rate of 12.5% year over year since 2017 and is projected to reach a revenue of US$3 billion in 2020.[39] In the prosthetic and dental industry, for instance, 3D printing has a vast scope. Many dental manufacturers increasingly depend on this technology to seamlessly design their products.

3D printing is a relatively new process in the market, and many people are unaware of this superior technology. It is based on an additive process that forms a 3D object through successive layering. These layers are thin and stacked together to take the shape of a tangible object.

Many organizations are exploring commercial 3D printing and adopting the technology to establish various prosthetic and dental industry products. The creation of 3D objects is vital in multiple fields as it is difficult to establish an item with such precision. The manufacturing cost is significantly reduced through the 3D printing technique, and it has significant potential to evolve manufacturing procedures.

The cybersecurity concerns specific to 3D printing primarily focus on connectivity and precision. Both connectivity and accuracy are integral to the overall process of 3D printing as any vulnerability or misalignment can carry safety and financial losses associated with the costs of incorrect printing. The increasing loophole and unprotected connectivity of a 3D printer within the network expose it to cyberattacks. Some threat actors may steal print designs, while others may hold the device hostage for a certain ransom amount.

Various organizations are looking for comprehensive solutions that can amplify the outcome of the 3D printing procedure. It is essential to tackle the baseline issues of 3D printing to make it a more viable industrial tool. Security experts are striving to make this technology more secure and an effective industrial instrument for better outcomes.

Cybersecurity experts have found a few viable solutions that make the process a lot more secure, such as creating backups and utilizing cloud solutions to ensure a seamless process. Organizations have also established centralized data storage for better inspection of data and implementation of data protection solutions. Installing configured software and regularly changing passwords has also proven as beneficial methods to enhance security.

Some security experts are skeptical about 3D printing technology due to its ability to hamper security operations. The biomedical industry and weaponry industry are experiencing the negative impact of 3D printing on a larger scale. In the biomedical sector, 3D printing can meet the need for donor shortages through rapid production of medical implants; however, existing security vulnerabilities, such as theft of intellectual property and production information, leave the engineering of print tissue organ implants at risk of sabotage and data loss. Adequate security and intellectual property protection solutions are the key to making the 3D printing process secure from potential cyberattacks and third-party interference.

What are the advantages of using 3D printing technology?

- **Fast prototyping:** Ideally, an effective prototyping procedure simulates the production process making it significantly faster. In the real world, a rapid change in manufacturing design can disrupt operations while also taking a lot of effort and investment. 3D printing technology is helping resolve this issue by building prototypes without interfering with the production line or requiring additional costs. The prototyping process is highly optimized due to on-demand 3D printing technology.

- **Production flexibility:** 3D printing technology is a viable production solution for various manufacturing organizations. With rapid prototyping, there is also a need for flexibility to quickly change designs to improve quality. This technology provides leverage to organizations to increase production flexibility to customize or modify products, which drives revenue.

- **Cost savings and minimum wastage:** Instead of fixed investment uncertainties in conventional manufacturing innovation methodologies, organizations can now realize cost savings with rapid concept development, instant feedback and faster time to market. 3D printing technology bears an initial capital investment followed by a sustained operational expenditure. However, the return on investment is quickly realized with product concept and development optimization. Furthermore, it is crucial for organizations to be socially and environmentally responsible and pay more attention to sustainability. 3D printing has minimum wastage, which helps organizations in meeting sustainability targets.

Where is 3D printing technology headed?

Commercial 3D technology is still in its infancy for global adoption. However, the trend is climbing as more and more industries require on-demand printing to meet mass production capacities faster. The data integrity and interconnectivity risks of 3D printing technology have given rise to various issues. With 3D printing forecasted to deepen its reach across multiple industries, adequate security controls like addressing intellectual property protection, privacy protection and product integrity must be integrated within 3D printing to make it a more widely adopted technology.

DATA PROTECTION

You may have come across the common phrase "data is the new oil." Like oil, raw data is not valuable in and of itself; however, value is created when the data is gathered entirely and accurately. With storage becoming more cost-effective and processing power in computing advancing over time, organizations these days can store, process and interlink massive amounts of valuable data sets.

As a result, organizations, states and countries have implemented privacy and compliance legislations to ensure organizations are appropriately safeguarding their critical and sensitive data by applying security measures such as the Confidentiality, Integrity and Availability (CIA) Triad. Data protection strategies are crucial in maintaining stakeholder trust and growth. Cybersecurity experts support organizations in meeting data protection compliance requirements as well as reporting potential uncertainties.

Threat actors target theft of sensitive data such as personal identifiable information (PII), personal health information (PHI), financial information, intellectual property (e.g., trade secrets, patents), and government or industrial secrets. Even a small vulnerability within an organization's people, process and technology can lead to data exploitation. A compromised security system can result in the loss of sensitive data costing the organization substantial fines and reputational damage.

According to the 2019 Global Threat Report from CrowdStrike, a sophisticated adversary needs less than 20 minutes to move laterally in an organization's network.[40] The report provides data about "breakout time," a CrowdStrike cyber metric that measures how much time it takes an adversary to move laterally within an organization's network from the point of breach. This insightful data provides organizations a critical benchmark about how quickly a data breach can occur, hence why it is vital to implement security strategies to mitigate current and potential vulnerabilities.

Strategies that include automated solutions such as anti-phishing and anti-malware software helps security experts understand rising uncertainties that can potentially target their organization. Data encryption is also a healthy practice most organizations continue to apply to protect data entering or leaving the network. Data encryption is a series of complex algorithms that protect the data even when it is stolen. The stolen encrypted data cannot be used as it is difficult to break through complex encryption algorithms. Data backup is an absolute necessity to ensure any attack, such as ransomware, on the primary data sets can be recovered through backups.

In addition to automated tools and data encryption, security experts also recommend insights from threat intelligence to better understand the operational capability of various attackers, implement and continually test incident response plans, and educate personnel regarding basic security principles. Effective maintenance of security initiatives and a well-informed workforce help meet organizational security goals faster. Organizations should invest an ample amount of time educating their employees about identifying vulnerabilities and actions to take to limit potential vulnerabilities.

What are the benefits of an efficient data protection program within an organization?

- **Better organizational performance:** Consumers, patients and the general public are sensitive about their data. They trust organizations to protect their data by complying with regulations and privacy state and federal laws. A benchmark for data protection regulation is the General Data Protection Regulation 2016/679 (GDPR), a regulation in European Union law on data protection and privacy. An organization with an adequate data protection policy tends to retain more stakeholder trust, which uplifts overall business performance.

- **Enhanced data security**: As threat actors continue to add new tools and techniques to their arsenals, an organization with weak data compliance may suffer data loss and valuable business relationships, causing long-term damage and limiting growth. Adopting comprehensive solutions is a viable approach in maintaining data authenticity and gaining a competitive edge.

- **Reduction in financial spending**: Organizations are always looking for better security solutions to reduce financial expenditure. However, insufficient data compliance can put organizations into a risky position where financial costs due to non-compliance can be high. Secure data protection strategies can prevent organizations from compromise while also save costs in non-compliance fines and potential lawsuits.

Where is data protection headed?

Immunity passport apps, contact tracing apps, and employee monitoring software are examples of technological innovation fraught with privacy risks and causing significant unease. The reason is security and privacy policies were an afterthought in the rushed development of these innovations.

The pandemic forced the travel industry, healthcare and businesses to rethink the safety of people while recalibrating how the world can continue to operate while in lockdown.

Immunity passport apps are designed to prove an individual has been vaccinated, allowing freedom to be out in public with limited restrictions; contact tracing apps are designed to track the movement of a potential infectious outbreak and notify people in close vicinity to limit the spread of the virus; and employee monitoring software sales surged to help employers monitor employee productivity from home.

Right now, the risks far outweigh the advantages of these innovations until information security and data protection are implemented the right way. For example, immunity passports enable access to patient health information (PHI) records such as test results, vaccine details, healthcare numbers and more. Then initial developments of contact tracing apps leveraged user mobile phone GPS or Bluetooth to determine proximity. And employee monitoring software tracks activities by taking screenshots, live video feeds, screen recording and keystroke logging. The risks of human safety, data theft and breach of privacy regulation are the reasons why these innovations have failed or are not as widely adopted.

Beyond the pandemic, there are technological advancements with microchip implants, passwordless access and digital identities. The digital future is here. With data as the new currency, the privacy and cybersecurity careers are far-reaching now more than ever before.

DevSecOps

In an ideal world, we would not be talking about security separate from software development. After all, we buy cars with airbags and locks already factory-fitted instead of over-laying them after the fact. Simply put, when security is baked into the DNA of the code our developers write, the time and cost to market is minimized. The current reality, however, is quite different.

With speed to market as the primary goal, security is often sacrificed. Organizations certainly do not want to deliver insecure products to their customers; they simply need security to be less of an inhibitor to speed and cost. Therefore, a set criterion for integrating security and compliance seamlessly into development life cycles is needed. The concept of Development, Security and Operations (DevSecOps) aims to solve this problem.

How do we solve this problem when our developers are not security experts? Well, this is an interesting question and an important one too. Addressing this conundrum requires a process, tool and awareness shift that seamlessly integrates security throughout the development and deployment life cycles.

Here are some tips that will ensure security becomes a partner and enabler rather than merely a tollgate:

- **Security and compliance testing integration:** Security needs to be a part of the developer's environment and not the other way around. Making the process complicated will not help the cause, and insecure code will continue to make its way to deployment cycles. Instead, the developer's environments should inherently have best practice integration and security testing tool sets, including secure code tips and tricks and a security peer-review process.

- **Skill set integration:** Breaking down barriers between development, security and operations teams can be challenging. Therefore, initial steps to remove these barriers may involve change management, integrating teams into the development life cycle, integrating skills through education and training, and leveraging partnerships through "security champions."

- **Security Assessment Center of Excellence:** Establishing a centralized assessment function for every build ensures risks are cataloged and activities, like security intelligence and secure development life cycle best practices, are integrated uniformly.

What are some of the advantages of incorporating an effective DevSecOps practice?

- **Cost savings through the early determination of vulnerabilities:** Addressing security flaws within a product at an early stage is much more cost-effective than addressing them after the fact. The proper implementation of DevSecOps technology can add to an organization's profitability as it helps determine potential vulnerabilities at the development stage. It is smart to assess software development vulnerabilities at their infancy stage to save on overall cost from reduced resource time and various liabilities.

- **Increase in security awareness:** Saving on operational cost and investment is critical for organizations for rapid progression. However, organizations may spend more to hire security experts to maintain security systems. While hiring security experts is of utmost necessity, training developers can reduce the cost of basic security rules. Reinforcing security controls at the development stage of software allows developers to be mindful of the vulnerabilities. This continuous security practice can improve basic security controls that can save organizations from security risks.

- **Reduce legal liabilities and risk:** Various software development organizations work on a contract with a third party to develop viable software. The risk is that few organizations validate the security of the software, and the lack of security controls can seriously damage an organization's brand reputation. DevSecOps can save an organization from legal liabilities and reduce the risk of reputation damage. However, the risk cannot be entirely mitigated, but it can be significantly reduced using DevSecOps security solutions.

Where is DevSecOps headed?

While the convergence between development, security and operations is happening around us, there are still substantial gaps in organizational maturity. This shift, however, definitely upholds the explicit underpinning of cybersecurity, which we have discussed throughout this book. From a cybersecurity career and skills point of view, DevSecOps is another area in which security expertise is required.

Threat Vulnerability Management (TVM)

Security vulnerabilities are weaknesses within a computing environment that a threat actor can exploit. Organizations continue to integrate vulnerability management into their security systems due to the global rise in vulnerabilities and an increase in third-party application deployment and compliance requirements.

Cyber incidents such as Heartbleed (2014), Meltdown and Spectre (2018) and Sunburst (2020) are examples of exposed vulnerabilities within the computing environment. What's more, approximately 60% of breaches are due to unpatched known vulnerabilities, according to a Ponemon Institute survey.[41] The continuous technological advancement within the IT field has made vulnerability management an essential IT operations requirement.

The purpose of a Threat Vulnerability Management (TVM) program is to prioritize and minimize the attack surface; this involves the proactive identification, evaluation, treatment and reporting of vulnerabilities existing in a system or organization. Treatments may include remediation, mitigation or acceptance of the identified vulnerabilities. A mature TVM program may consist of conducting regular and comprehensive scans; continually assessing vulnerabilities across environments; accelerating processes with automation tools; closing prioritized security gaps promptly; and facilitating regular and customized security training for various departments.

The process of vulnerability management involves configuration change implementation, enabling vulnerability monitoring services and blocking exploitation through Intrusion Prevention System (IPS) devices. The vulnerability management process is an old concept; however, organizations are consistently on the lookout for new TVM technology to effectively combat any vulnerabilities within their systems. Updated and improved vulnerability systems give organizations a sense of confidence to deal with uncertain situations effectively.

Vulnerability management is a combination of processes and tools, such as scanning and patch management. There is an expected increase in the adoption of vulnerability management as organizations mature their IT operations, including efficient patch management and reporting. Patching operating systems, applications and network equipment helps minimize vulnerabilities and threats to avoid potential threats and attacks. TVM reports help to determine patterns and trends of an organization's vulnerabilities and help inform the decision-making process to remediate, mitigate or accept those vulnerabilities.

What are some of the advantages of having a well-functioning Threat Vulnerability Management (TVM) program?

- **Enhance organizational security:** The main reason an organization looks to adopt vulnerability management is to improve its security hygiene and posture. Vulnerability management is a valuable method for inventory management, patch management, vulnerability scanning and risk assessment. More and more organizations of all sizes choose to outsource vulnerability management services to the scale and scope tailored to their specific business need.

- **Time-efficiency:** Through automated vulnerability management tools and services, security experts can quickly determine the existing and current vulnerabilities as well as establish a quick response for remediation. Consequently, more resource time can be allotted to tasks that cater to organizational growth.

- **A holistic view of risks:** A well-implemented vulnerability management program provides an organization with a holistic view of risks as it monitors the system activity continuously. The information gathered from vulnerability assessments, for instance, helps to contextualize the data and help inform decisions to combat certain security weaknesses. Regular assessments, monitoring and reporting help an organization proactively stay on top of the state of vulnerabilities and remediate accordingly.

- **Cost savings:** It can be expensive and challenging for organizations to manage vulnerability management themselves. Finding and retaining talent is a prime example of a challenge. Also, for small businesses, vulnerability management often lies with the IT team that may experience challenges, such as the lack of expertise about emerging threats; the inability to prioritize those threats effectively; and the lack of an adequate governance model that aligns with the overall risk tolerance of the business. By outsourcing, an organization has the assurance the right expertise, tools and reporting are available and implemented. Vulnerability management tools and services are also tailored to the size, scope and budget of an organization. In turn, the organization maintains the visibility of existing and potential security weaknesses, remediates as needed and saves costs over time.

- **Regulatory compliance:** Vulnerability management plays a critical role in compliance efforts specific to data protection under GDPR, HIPAA and PCI-DSS requirements. To maintain compliance, rigorous practices for data protection under vulnerability management include asset inventory, risk assessments, continuous vulnerability monitoring, patch management and remediation.

- **Evaluate third-party performance:** Organizations are increasingly dependent on outsourcing security services from a third party such as system administration, email, backup and VoIP, to name a few. Sometimes the security risk can be unintentionally imposed by the third-party service provider, which remains unknown until organizations have their independent assessment for vulnerability assessment. Adopting vulnerability management can enhance an organization's ability to evaluate the performance of a third party and determine its reliability.

- **Remediation efficiency:** Not every vulnerability is relevant to an organization. Therefore, a TVM program designed with a targeted risk-based approach helps focus remediation efforts on the most likely threats, vulnerabilities and risks. Organizations that have seen success in their vulnerability management programs start with a current state benchmark and improve over time by tracking metrics. Key performance metrics may include the number of vulnerabilities remediated, measuring the precision of remediation and time to remediate based on criticality, and improving capacity ratios (e.g., for every five vulnerabilities open, one is remediated within a specific timeframe). Establishing a benchmark with performance metrics helps guide improvement efforts to reach desired security maturity objectives.

Where is TVM headed?

Threat vulnerability management may seem like the ultimate solution to all the vulnerabilities arising in the security system. However, it too needs continuous improvement and better tools and processes for complex networks within organizations. The pain point for many organizations is understanding how to prioritize vulnerabilities in a time and cost-efficient way. Therefore, organizations are moving TVM toward a more risk-based approach to prioritizing remediating vulnerabilities based on risk criticality in alignment with organizational risk tolerance. Gartner identifies risk-based vulnerability management (RBVM) as the #2 priority in its *Top 10 Security Projects for 2020-2021* eBook published September 2020.

Furthermore, automation in RBVM will reduce manual tasks and improve automation analysis, exploit prediction, network and assets mapping, risk assessments, prioritization, remediation and much more. Demand for cybersecurity professionals within TVM will also continue to grow to help organizations build and manage their TVM programs.

SOFTWARE-DEFINED (SD) SECURITY

Gone are the days when traditional IT infrastructures were sustainable and viable. The pandemic catapulted companies to realize the necessity of a scalable and flexible computing environment. It was just over a year ago when the world went into lockdown, and cloud computing and virtualized infrastructures became a priority for business networks.

That is where Software-Defined Security (SDSec or SDS) solutions and the process of virtualization play a critical role in the way companies meet the need for a secure and remote workforce. SDSec is when a computing environment is controlled and managed by security software. It covers a range of capabilities, including intrusion prevention and detection, firewall, passive and active scanning, device and user identification, network segmentation, threat prevention, application controls, decryption and more.

The SDSec model also enhances the network function through Software-Defined Networking (SDN). It contributes to establishing new security layers and lays a firm ground for an organization's network security central management, optimal usage of bandwidth and compliance throughout an organization's network components.

What are some of the advantages of using SDSec technology?

- **Intelligent hardware integration with software:** SDSec allows organizations to implement various security controls using software, such as device and user identification, threat prevention, application controls, decryption and more. SDSec is beneficial because of automation, abstraction, scaling, Application Programming Interfaces (APIs) and orchestration.

- **Centralized control:** An environment with virtualized security enables an expert to carry out managerial tasks through centralized control remotely. The benefits of centralization with SDSec include a holistic view of security policies across virtual environments and integrated hardware and software configuration coordination. In turn, these benefits accelerate threat detection and consistent policy delivery to help prevent security risks in the shift to digitalization.

- **Reduction of hardware cost:** The reduction of cost can act as a primary beneficial factor of SDSec. The virtualization of security network applications and operations within the hardware commodity cuts down on an organization's need to deploy or buy specialized vendor appliances. An organization can adequately utilize this cost to cater to other business areas requiring more investment.

- **Information visibility:** SDSec is a comprehensive virtual solution that involves intelligent detection and prevention across resources through the collaboration of SDSec and the SD network. How it works is an element of network probes across various locations within the network; then, whenever the system detects any vulnerability, it utilizes past information and correlates it to produce a response. With the increase of data visibility, results may include a reduction in Mean Time to Detect (MTTD) and Mean Time to Recovery (MTTR) metrics.

Where is SDSec headed?

Before the global pandemic, some organizations placed SDSec on their digital transformation roadmaps. Many more organizations since the pandemic have started adopting the SDSec model and virtualized technologies to meet the need for a secure remote workforce through programmability, scalability, automation and policy-driven security architectures. Virtualized infrastructures also help organizations improve data loss prevention through intelligent management of security threats and attacks.

To deploy SDSec, an ample amount of expertise in networking, engineering and IT architecture is required for seamless implementation, including the modernization of governance, policy controls, security controls and risk management. The need for cybersecurity professionals will continue to grow as organizations enhance their SDSec portfolios.

Security Orchestration Automation and Response (SOAR)

When an organization faces a potentially threatening security event, it usually triggers a manual incident response journey from mobilizing teams through triage, investigation, remediation, reporting and so on. This manual process increases the Mean Time to Recovery (MTTR) metric.

CrowdStrike suggests organizations aim to achieve the 1-10-60 rule for cyber incident response: 1 minute to detect, 10 minutes to investigate and 60 minutes to remediate. To reduce the MTTR and achieve the 1-10-60 rule, automation and little-to-no human involvement are essential. Security Orchestration Automation and Response (SOAR) helps with this endeavor.

SOAR helps orchestrate and automate threat detection and incident response workflows. It achieves this by integrating tools, systems and applications used by SecOps in the Security Operation Center (SOC). Through SOAR tools, SecOps can concentrate more on critical tasks as opposed to managing numerous threats manually.

Highly adaptable and flexible, the SOAR tools and technology can integrate into the security framework of most organizations. There are three SOAR technologies, including threat and vulnerability management, security incident response and security operations automation. SOAR technology often complements a mature Security Information and Event Management (SIEM) platform. Organizations implement both SIEM and SOAR technologies to enhance their command over threat and vulnerability management.

What are the advantages of using SOAR technologies?

- **Time and cost savings:** With automation comes time and cost savings over time. When properly configured and in alignment with a risk-based security strategy, the return on investment of SOAR technology makes it all the more beneficial. SOAR technologies assist organizations in managing several cybersecurity incidents in an automated and predictable way, streamlining the incident management processes and making it highly efficient for time and cost.

- **Automation of routine tasks:** The capabilities of SOAR technologies enables an organization to automate routine and time-consuming tasks. With time savings from automated solutions like SOAR, organizations can utilize their existing experts to focus on strategic priorities such as planning and implementing proactive security measures.

- **Proper management of incidents:** SOAR technology integration allows organizations to rapidly respond to events and determine a solution for on-time remediation through automation. By eliminating the risk of human errors, SOAR technology enables proper incident management through accurate, consistent, intelligence-driven data, processes and workflows.

Where is SOAR headed?

More and more organizations are becoming aware of SOAR technology and its benefits to manage sophisticated cybersecurity threats. As such, SOAR technology is rapidly becoming the leading technology integrated by organizations to enhance the overall efficiency of security operations, primarily within their SOC environment.

Not surprisingly, SIEM solutions are also developing their SOAR capabilities as well as acquiring SOAR solutions. To that end, SIEM solutions are vastly competing with SOAR solutions offering increased flexibility in cost options for businesses of all sizes and budgets. As mentioned earlier, SOAR technology and SIEM platforms highly complement each other. The new wave of integrated solutions is creating increased opportunities for businesses to improve their security operations capabilities.

As the adoption of SOAR technology continues to rise, the need for cybersecurity practitioners with blended knowledge of SOAR and SIEM solutions will also continue to grow.

CHAPTER 6 KEY TAKEAWAYS

- Due to the COVID-19 crisis, security and IT teams suddenly had to support a remote workforce almost overnight. Securing remote employees across a highly distributed infrastructure was a massive undertaking for many businesses globally. Endpoint protection heightened identity and access management policies and solutions (e.g., two-factor authentication 2FA], Virtual Private Network [VPN]) and increase in employee security awareness training was just a few of the ways businesses needed to respond to the crisis.

- Additional trends at the forefront of cybersecurity today include the leap to the cloud; advancing and automating security solutions through artificial intelligence and machine learning; securing devices and equipment in the Internet of Things (IoT) and Operational Technology (OT); protecting innovation in robotics, connected and autonomous vehicles, blockchain and much more.

- Advancements in these technological trends prompt organizations to hire cybersecurity experts to help protect their business—people, process and technology—from security threats, attacks and reputational damage.

Chapter 7: How do I get a job in cybersecurity?

Now for the million-dollar question: How do I get a job in cybersecurity?

This chapter is dedicated to your pressing cybersecurity career questions. Sometimes these questions keep you up at night because answering "What do I want to be when I grow up?" can be downright challenging, adding stress, anxiety or both. Sometimes these questions are top of mind as you browse job boards after a frustrating workday. Sometimes these questions nag at you because you are not doing what you genuinely want to do; you feel discouraged because it feels like an impossible goal; or you are simply afraid to make a change. Or possibly, you are learning how talented you are and know that cybersecurity is the right career for you.

The time, money and commitment to set yourself apart and embark on a rewarding journey is no laughing matter. Some of you are already struggling with work-life convergence from remote work. Some of you recently lost your jobs and are struggling to make ends meet yet want to make a change. Some of you work the night shift only to go home to young children and a busy household with minimal time to rest. Some of you are single parents trying to create a better future for your family and you. Some of you are eager and ready yet may have little to no funding available.

We absolutely understand your situation. We have been in your shoes – blood, sweat, tears, all of it. Is it worth it? Unanimously, we all agree it is challenging and absolutely rewarding.

First, let's celebrate you:

- For all your hard work to date juggling school or work and home;

- For not giving up and instead re-imagining the possibilities of a future in cybersecurity; and

- For reading this book where you may begin devising a plan of what you need to do next.

Now, let's get to your questions.

How do I get started in cybersecurity?

Understandably, starting or transitioning into a career in cybersecurity can seem intimidating at first. In a highly competitive job market, you must make wise choices when it comes to career development. Cybersecurity is a fast-moving and fast-growing profession providing a lot of benefits to practitioners. People and organizations widely acknowledge the value of this field, and the employment gap makes it even more advantageous.

The cybersecurity field needs you!

Here we offer our recommended steps to kick-start your career in cybersecurity by introducing the **SIM method**: select your cybersecurity specialization, interview a cybersecurity practitioner and map your career.

1. Select your cybersecurity specialization

To gain expertise in the field of cybersecurity, you must first evaluate certain factors. Cybersecurity is a distinct field, and each organization has diverse demands and security issues. Before opting for cybersecurity as a field, do your complete research and determine your area of interest leveraging the CISSP CBK domains (available in Chapter 3) as your launching pad.

Firstly, determine if you are interested and passionate about a cybersecurity career. From there, you must choose a specialization, such as security administration, vulnerability management, security operations, penetration testing, risk management, security audits and so on.

While selecting your area of choice, consider if you have existing work experience that can transfer into your preferred specialization and if you need more education or training to enhance those skills further. If further education is required, an increasing number of knowledge resources and institutions are widely available to help you become a professional cybersecurity practitioner within one to two years.

- **DO NOT** fall into a rabbit hole with your research. Save yourself wasted time and unnecessary stress by starting your research with a purpose.

- **DO** use the CISSP CBK domains and your existing work experience and education as your compass in selecting your areas of cybersecurity specialization.

- **DO** start on the right foot, check your ego at the door and keep your eye on the prize. Accept and commit to the hard work to begin your career. It will be worth it, absolutely.

- **DO** visit cyberseek.org and explore the interactive cybersecurity career paths and the recommended knowledge, skills, abilities and tasks required to succeed in those roles. This interactive tool leverages the NIST NICE Framework discussed in Chapter 4.

2. Interview a cybersecurity practitioner

Take charge of your career journey mapping by speaking with one or some cybersecurity practitioners or reach out to someone who can refer you to a practitioner. It is beneficial to talk with a practitioner who works in the cybersecurity domain(s) you wish to pursue.

Be professional and purposeful when setting up your interviews. Provide the professional with a brief (keyword here is "brief") and concise introduction about your interest in cybersecurity as a career, your intent to glean insights about the profession and learn what the professional can impart as you begin your journey. Do not write a long letter. Also, practice courtesy by confirming the professional's interest and availability to speak with you.

Once you've secured a date and time, preparation is critical! Show the professional initiative by preparing relevant questions in advance. She or he will want to learn about your education, work history and passion. Likewise, you will gain her or his insight and recommendations for entering the field.

- **DO NOT** expect or ask for a job when interviewing practitioners; likewise, do not expect or ask for a job when connecting with cybersecurity professionals on social media channels.

- **DO** appreciate the time cybersecurity professionals take to speak with you by expressing gratitude. Be mindful they could have deadlines or time restraints, yet they choose to help you as you navigate your career journey. In the future, pay it forward when it is your turn to mentor and coach emerging cybersecurity practitioners.

3. Map your career journey

Once you have selected your preferred specialization and interviewed a cybersecurity practitioner (or some), you are ready to begin mapping your career progression. By this point, you should be able to determine the education and training you need, the certificates you will pursue, the cybersecurity associations you will volunteer with, and the entry-level roles you will pursue. Your career journey map should provide you a clear path forward, including the amount of time and budget you anticipate to achieve specific milestones.

It is also essential to be realistic and practical with the amount of time it will take you to achieve those milestones. For example, the job search process can take anywhere from three months to a year depending on various factors, including your geography, demand for entry-level positions in your desired specialization, your resume and more; this can undoubtedly be one of the more frustrating challenges of starting a new career. We explore this more under "What if I don't get a job interview?"

Remember, there are many who are rooting for you to succeed. Be proactive and seek out a cybersecurity mentor who can provide guidance in your career journey mapping and provide additional insights along the way.

- **DO NOT** skip this important step. You don't need anything fancy to capture your career journey. Research examples, then write it down, draw it or journal it. It's up to you!

- **DO** expect your plans to change because life happens. After you've created your career journey, save it, revisit it and make edits as necessary every three to six months. If there are deviations, make sure to return to it as part of your commitment to yourself in achieving great things in cybersecurity.

This **SIM method** provides a cornerstone approach on the three most important aspects of getting your foot through the door: select your cybersecurity specialization, interview a cybersecurity practitioner and map your career journey.

WHAT ARE THE COMMON MISTAKES I SHOULD AVOID WHEN CHOOSING TO START A CAREER IN CYBERSECURITY?

Every career has a unique trajectory based on the choices behind its evolution – some choices might not be suitable for others, but all choices teach something important. As Albert Einstein said, "A person who never made a mistake never tried anything new." However, there are certain situations that one can avoid early on in their careers to invoke the "fail-fast" rule. The fail-fast rule is based on the philosophy of extensive testing and elimination to determine if an idea has value.

As explained in the previous sections, there are certain pitfalls you can avoid when choosing to start your cybersecurity career journey, including:

- **Thinking that cybersecurity is highly technical and hands-on only:** As discussed extensively throughout this book, although cybersecurity is about security technology, it is not all about eyes on glass and fingers on keyboards. There are non-technical and technical cybersecurity work roles explored in detail in Chapter. 4.

- **Cherry-picking based on salary alone:** This is one of the most frequently made career mistakes (not just in cybersecurity). Focusing on salary alone results in getting blindsided to other aspects of a role: workplace culture, growth trajectory, learning opportunities and work-life integration, to name a few. Although what you bring home is essential, getting fixated on one number alone is detrimental. Increasingly, organizations are building total compensation packages that go beyond just the annual take-home pay. Carefully scrutinize what is vital should be considered when making the final decision.

- **Getting stuck on job titles:** Although job titles hold value and individuals attach pride to them, getting too picky about job titles could also be detrimental, especially when starting afresh. Not all organizations are equitable when it comes to job titles. In this scenario, a "bigger picture" view should be considered, and understanding what else is to be gained is more important.

- **Rushing the process:** For those starting a career in cybersecurity or transitioning mid-career, it is important to map out your career journey and understand that your career in cybersecurity requires the right education, the right certification and the right experience to reach your goals. When mapping out your career journey, include realistic time frames to complete certain stages, such as education and the job search. Allow time for life events as change is commonplace, and recalibrate your timeline as needed. This process, all in all, takes time and is not achieved overnight. However, when you do it right from the start, you can expect your career in cybersecurity to flourish.

- **Disregarding the value of your name:** In cybersecurity, most especially, your name, your reputation and the integrity of your background matters. Employers take extra care when hiring cybersecurity practitioners through background checks and open source intelligence, including social media across various potential aliases if applicable. If you are required to obtain a security clearance, you can expect a fairly extensive review of your entire work history, your education spanning your lifetime, your financial background, your family's background and more. What you think can't be seen or discovered is fair game. Honesty is the best policy when it comes to your career in cybersecurity, from your social media presence to your resume, job interviews and career progression. You have one reputation so take care of it.

- **Getting cozy:** It is far too common to find a job that one can cruise along in without much challenge or change. Although these situations are far too common, breaking away from them is sometimes required to see what lies beyond the horizon. Venturing out of your comfort zone requires some calculated planning and risk-taking. Of course, for anything to change, the status quo needs to be challenged.

WHICH EDUCATION PATH DO I CHOOSE TO BECOME A CYBERSECURITY PROFESSIONAL?

Education is a fundamental building block for acquiring proper knowledge within the field. With changing times, cybersecurity is rapidly progressing as a top career choice, and various organizations are desperate to hire well-informed security professionals.

However, the right education path is dependent on where you are in your life. You can choose between a certificate program, an undergraduate program, a master's program and more. Since post-secondary education is not free in most countries, it is important to take the time to evaluate your options while keeping cost and time top of mind. Perhaps you are early in your career, and a four-year bachelor's degree is right for you. If you are a seasoned professional, you may leverage your existing experience and consider a part-time cybersecurity continuing education studies program to obtain the proper knowledge base about the field.

Some organizations frame educational criteria for their cybersecurity professionals. Some employers recommend their employees gain a master's degree, for example, which gives cybersecurity professionals command to meet the organization's needs while increasing credibility. Once you are a cybersecurity professional, it is worthwhile exploring these education programs with your company.

Supplementing your cybersecurity education and work experience with the right certifications adds to your qualifications as a desired cybersecurity practitioner in the field. You can earn basic and professional-level certifications to boost your skills and expand your expertise. A master's degree or Ph.D. level education is optional based on your career goals but is not an absolute requirement to succeed in a career in cybersecurity.

IF I DON'T HAVE TECHNICAL EXPERIENCE, HOW DO I GET A JOB?

Oftentimes, there is a perception that to work in cybersecurity, you need to be highly technical in areas like penetration testing, network security, coding and infrastructure. Although that would be true for particular roles, there are many non-technical roles within cybersecurity.

In Chapter 4, various cybersecurity career options are discussed in detail, and there are a number of non-technical roles that do not require an extensive technical background or training. These roles include cybersecurity risk management specialist, cybersecurity legal advisor, cybersecurity auditor and cybersecurity sales and marketing specialist, to name a few.

A cybersecurity sales and marketing specialist role, for example, would require a moderate understanding of the technology you are selling. After all, your clients will ask you questions about how the product you are selling can integrate within their organization and help protect their critical data.

An example of a highly technical role discussed in Chapter 4 is that of a Penetration Tester. These roles require a precise understanding of areas such as communication protocols and technologies, hacking tools such as Kali Linux, Command Line Interface (CLI), knowledge of operating systems services and registries, being able to write scripts or code and the ability to write management-level reports (i.e. soft skills).

For those of you with no technical background, leverage your existing work experience, research the right post-secondary institutions offering what you are looking for, register for a cybersecurity program that fits your objective and life-style, excel in your courses, complete your education. As you near the completion of your education, begin applying for cybersecurity positions that complement your work experience and cybersecurity education.

For those of you who are a high school student, an undergraduate student or a continuing education student, leverage information from this book and seek guidance from your teachers, professors, instructors or career center for additional support.

What are examples of entry-level, intermediate and advanced cybersecurity positions?

Role levels vary by organization. Building on Table 1 from Chapter 4, here is a general summary of entry, intermediate and advanced cybersecurity positions. Some of these roles may also start as IT roles with the opportunity to grow into cybersecurity specialization. These roles are some of the keywords you can research as part of your job search. Additional suggestions are available on CyberSeek.org, an interactive cybersecurity career pathway tool created leveraging the NIST NICE Cybersecurity Workforce Framework also mentioned in Chapter 4. These roles are suggested examples for reference only.

Table 2

Entry-Level Roles	Intermediate to Advanced Level Roles
Security Analyst	Security Engineer
Security Operations Analyst	Incident Responder/Handler
IT Support Specialist	Penetration Tester
Network Support Specialist	Computer Forensics Expert
IT Project Coordinator	Malware Specialist
System Administrator	Threat Hunter
Vulnerability Analyst	Information Systems Software Developer
Cybersecurity Sales and Marketing	Cybersecurity Legal Advisor
Security Consultant	Privacy Officer
	Program or Project Manager
	Security Assessor
	Cybersecurity Auditor
	Cybersecurity Instructor or Trainer

	Information Security Architect
	Information Security Manager
	Cyber Risk Management Professional
	Chief Information Security Officer (CISO)

It seems most cybersecurity job postings require a minimum of 5+ years of experience. Am I then not qualified?

Job postings, as is today, need improvement. You're right – job postings ask for the world and quickly discourage prospective talent. Work is in progress as more cybersecurity experts collaborate with recruiters to create job postings that attract a diverse talent pool.

Until job postings improve across the board, we encourage you to know which domains in cybersecurity you are interested in pursuing, couple that with your work experience and apply to as many job postings as possible. The best that can happen is you get a job interview, right?

It is also helpful to know hiring managers and recruiters are acutely aware the perfect candidate does not always exist. As such, they are flexible when reviewing cover letters and resumes to find a handful of candidates who meet some or most of the qualifications they seek while also understanding there may be some additional training required.

Of course, it is recommended to apply for cybersecurity job postings that reasonably complement your existing skill sets and knowledge base. Be honest with your self-assessment of what you can do on the job. Expect that interviewers will ask you questions related to the responsibilities of the role, so you must know exactly what you can do, have the fundamental knowledge to carry out the work, be honest if you do not have specific experience (because this is okay to say!) and have confidence in the value you bring to the organization.

You'll also be pleasantly surprised to learn that some hiring managers and recruiters will choose a candidate who may only be 60% qualified. They are willing to train the candidate because she or he is a great fit for the organization and shows enthusiasm, positive energy and passion, which are personality characteristics valued more often than alphabets that follow a name.

Even if you are 60% qualified, apply. Apply. Apply. Apply. Cybersecurity needs you!

I'M A SEASONED PROFESSIONAL AND LEADER IN MY CURRENT ROLE. WILL I HAVE TO START AT AN ENTRY-LEVEL POSITION IF I TRANSITION TO CYBERSECURITY?

The honest answer is "it depends." For example, if you are already in IT management, you may have to make an intermediate lateral move into cybersecurity. You can consider negotiating with your employer your business case for maintaining your salary as is (e.g., tenure, excellent performance, etc.) while you train and learn on the job within the new cybersecurity role. Be prepared that this conversation may not go the way you want it to because positions have a salary cap. It is worth exploring no less. Some small-to-medium businesses may have more flexibility in making this work.

If, for example, you are not in IT management and new to cyber and IT altogether, you may have to sacrifice a few years at an entry-level to an intermediate role before you can realize opportunities for cybersecurity management. You may find you are assigned more than your fair share of junior work than preferred. It is highly advisable to design your career path to align with all those you work with at your existing or new company. Ensure you clearly and regularly state your career goals, your value add and align with them on a reasonable timeline. Above all, remind them that you are a seasoned professional who wants to start from experience, not from scratch.

It may take some time to find the right company that will support you 100% in your career development as a seasoned professional. The right company will utilize the skills you bring, will assign you the right work to help with your career plan and ensure you are gaining the right cybersecurity experience.

You may not find the right company right away. Be patient with the process, be kind to yourself, trust in the expertise you bring to the table, be relentless in believing in yourself, be professionally vocal and continue to pursue it with fervent passion. When you don't achieve what you want within a certain period of time (e.g., two years at a company), start your job search again and move on to a position where you will thrive.

Is it better to transition into cybersecurity within my existing company?

There is power in leverage. It is easier to make a lateral move into cybersecurity through your existing employer, where you already have your foot in the door, than to start as an external hire with a new company. In this case, leveraging by way of gateway opportunities within cybersecurity at your current workplace is recommended. Gateway opportunities are jobs requiring you to help with cybersecurity projects without you having to be the subject matter expert.

For example, a cybersecurity project manager, project coordinator or business analyst are roles that help with project execution and coordination; this might be a great starting point to enter into the field at your existing workplace. These roles would allow you to work closely with cybersecurity subject matter experts and begin learning their lingo.

The good news is most IT roles nowadays likely have a component of cybersecurity, directly or indirectly. It makes a lot of sense to find out about these roles are internally at your organization. Stay close to job postings within your organization and reach out to the hiring manager and recruiters to find appropriate details about the role and opportunities for cybersecurity cross-pollination. A quick search on Indeed.com reveals the following roles that fall in the above gateway categories:

- Business Analyst - IT Security: This role works closely with security architects and consultants to capture security-specific requirements in applications.

- Project Coordinator, Security Program: This role coordinates project activities, including critical path management, stakeholder communication, document repository management for all cybersecurity projects within the enterprise Project Management Office (PMO).

- Project Manager, Cybersecurity: This role is responsible for all aspects of the smooth delivery of cybersecurity projects, including managing project budgets, timelines and relationships throughout various lines of business.

Clearly, there are roles within your organization right now that will enable you to work directly with cybersecurity subject matter experts, thereby building the right foundation for you to land a job in cybersecurity eventually.

Do I Need a Degree or Diploma?

In this specific case, the most relevant answer is "it depends." According to Glassdoor, some companies no longer require a degree.[42] But there is a caveat: this is the case for a small fraction of jobs (not all), and it depends on the country you reside. Read the fine print. For example, Google made it to this Glassdoor list but not for all roles.

Here are the roles that, according to this article, Google no longer requires a degree for: Network Specialist, Software Engineer, Associate Contracts Manager, Revenue Lead, Head of Sales Knowledge Management, Digital Marketing Lead for Google Fiber, Hardware Engineering Intern, Business Intelligence Manager, Senior Interaction Designer, Account Strategist, and Technical Program Manager. If you look carefully, most of these roles require very specific experience that, in lieu of a degree, needs to be balanced with specific hands-on experience. There is an increasing number of companies willing to interview people without a traditional degree.

However, most intermediate-to-advanced roles in cybersecurity continue to require a degree or at least a college diploma. While it is possible to find entry-level and specific technology roles within cybersecurity without a formal degree, most jobs still require a four-year degree in cybersecurity or related fields such as computer science or technology engineering. A formal degree prepares you for various other things apart from just the subject matter; these include soft skills such as time management, communication skills and problem-solving. These are life skills that remain pertinent no matter where you end up.

When you apply for a cyber job (or any job for that matter), you would likely be up against several candidates with similar or better profiles, and most will have a degree. Employers look at candidate profiles holistically including education, experience, certifications, gaps in employment and so on. However, if you lack in all the above areas, the chances of you getting your foot in are drastically reduced.

Let us be honest, the competition is real. While thinking of building your profile, think about creating that right balance. If, for example, you have no relevant experience, an employer might look at relevant formal education to get you in at an entry-level, so you can apply your learnings at the job and excel with time, thereby making you very desirable.

A quick search on Indeed.com for an IT security analyst role, which is mostly entry-level or intermediate, reveals that employers require a college or university degree/diploma in a "relevant" field, with some strongly preferring a cybersecurity specialization.

To summarize, the chances of landing or advancing in a cybersecurity role are strongly amplified with some form of formal education in a college or university setting, especially if one is weak in other areas that employers desire (e.g., relevant experience).

WHAT IS THE VALUE OF CERTIFICATIONS?

This is a very common question we are asked as educators and hiring managers. The answer to this question is a little tricky, as it presents the proverbial chicken-and-egg situation. One thing is clear, certifications on a resume increase your potential to get an interview and to advance in your career.

Chapter 5 outlined in detail the various cybersecurity certifications and the respective salary potential. Increasingly, employers are moving certification requirements from the nice-to-have section to the mandatory section. The likely reason behind this transition is that security professional requirements and skill set inventory has become more streamlined and formalized (thanks to regulations and best practice frameworks). Professional certifications generally map their body of knowledge to these regulations and best practices.

Do certifications replace experience and vice versa? Not really. However, what makes a candidate competitive and desirable for employment is the right combination of both. Understandably, there must be a starting point where an individual would be light on both (experience and certifications). In that case, the expectation needs to be realigned, including identifying where one starts at and this would likely be an entry-level or co-op opportunity.

For experienced professionals, your existing employer may promote and support your career development in acquiring certifications to ensure your role continues to bring increased value to the organization. First, you must determine which cybersecurity role you want long-term and understand which relevant certificates are right for you to achieve it. Take this information and ask your employer to support you in your growth and development endeavor to become a certified cybersecurity practitioner.

WHY DO I NEED TO NETWORK?

A job search may be easier if you "know" the right people. This is proven time and time again. When you are new to cybersecurity (or to any field), smart and effective networking helps you increase your chances of getting your foot in the door.

Your strongest artillery in a job search is the power of professional referrals. As people in the field become familiar with you as a valuable asset, the stronger your brand profile becomes and the greater the chance of being shortlisted for an interview resulting from the "people you know."

To some, networking is a daunting task, especially if you consider yourself an introvert. One of the interesting developments of the COVID-19 pandemic is the rise in social anxiety among those who did not have social anxiety before. If this applies to you, sometimes networking is easier when you start with those you know first. From there, network outside of your circle in bite-size chunks.

Do you have a list of at least five individuals you can call to brainstorm ideas and receive career advice? For this book, we call this group the **Cyber5 Hotline**. If you have not yet developed your Cyber5 Hotline, now is a great time to start building that. To get started, think of everyone you interact with (both professionally and personally). Now think of who out of those can give you valuable professional advice. The fact is that if you never ask, the answer is always a "no." People are generally helpful, and, in some cases, just a five-minute conversation sometimes might mean a leap in career advancement.

So where are these people who would make your Cyber5 Hotline? This group might consist of individuals from your family, coworkers, hobbies, clients, customers and more. Think through that list and reach out to people who you think can be a member of your Cyber5 Hotline.

There are various things your Cyber5 Hotline might help you with, including resume feedback, spreading the word about you and helping you prepare for an interview, to name a few. Building and calling upon your Cyber5 Hotline can help you identify the most effective channel to job hunting, which in turn can save you time, effort and stress.

When thinking of starting to network and building your Cyber5 Hotline, you should start inside your workplace. This is not only the easiest place to start but also lets you leverage existing relationships. For example, if you work at a bank and are looking to pivot into cybersecurity, reach out to a cyber-security hiring manager or send a quick note expressing your interest in the field. Ensure you backup your outreach with some tangible steps you have already taken. Do not make it sound like a cold call. Put an extensive effort into knowing all that there is to know about the department and how you can bring value.

Before you reach out, ensure that you have already taken a couple of steps in that direction. For example, enrolling in a cybersecurity course and sharing your one-year career jour-ney plan in the first question of this chapter. A great idea would be to reach out to a coworker in that department and request if shadowing or on-the-job training a couple of hours a week can be arranged. Take your coworker out for lunch if you have to, or send them a Starbucks gift card (as an exam-ple) as your token of appreciation.

Find coworkers in the break room or send an instant message. Ensure you do some homework and learn about their career progression on LinkedIn. Ask questions about how they built their career in cybersecurity. You would be surprised to see how many people are willing to share their career lessons learned.

Here are some ideas to help with networking:

1. Join professional networking groups, such as LinkedIn groups. Some examples include the Information Security Careers Network (ISCN), local (ISC)^2 chapters, cybersecurity job network, Information Systems Security Association (ISSA) Discussion Forum, InfoSec Careers (Information Security), ISO 27000 for information security management and many more.

2. Network at local security group meetings and networking events. For example, within the Greater Toronto Area (GTA) in Canada: (ISC)^2 Toronto Chapter; TASK - Toronto Area Security Klatch; OWASP Toronto Chapter; Cyber Tech & Risk; Leading Cyber Ladies Toronto; Women CyberSecurity Society - Toronto, ON; Toronto Cybersecurity for Control Systems. Search for local groups where you live and find out how you can be a part of it. There are also volunteering opportunities in many of these groups, which might be a great entry point.

3. Volunteer for a cybersecurity project at school or work. Everyone can use some help with projects, so reaching out to classmates or colleagues to offer help with project tasks might be a great idea.

WHAT CAN A MENTOR DO FOR ME?

Simply put, a mentor is a trusted advisor who is willing to share her or his advice and is happy to provide guidance and wisdom to you. Again, the best place to find a mentor is at your workplace. If you are right out of school or in-between jobs, you can ask a teacher or someone within the cybersecurity profession via a LinkedIn search. We receive many requests through LinkedIn where aspiring cyber professionals request to speak to us as their mentors. Although cold calling does not always work, some do go out of their way to help.

Some organizations have a formal mentorship program. If your organization does not, you will have to be extra proactive. One thing to remember here is that your career is in your hands, you are in the driver's seat, and a mentor or your network will only provide valuable guidance on the path you have chosen to embark on. When interacting with your mentors and your network, never make extra work for them, otherwise, you will soon be dropped and as they say, you will be "left on read."

A mentor is someone who has specific, insightful knowledge that she or he may have been exposed to over time and are now willing to share with you. A mentor can also sometimes help introduce you to certain key people within her or his network. When identifying people to be your mentors, spend some time in knowing them by researching their career trajectory and asking them specific questions such as their career path, education and certifications, benefits and lessons learned.

Approaching someone to be your mentor can be tricky. A good approach might be to have someone broker the initial conversation; this could be your boss or a coworker. Once the introduction has been made, ensure you tell them you want them to be your mentor and why you feel the need for this type of relationship.

Through the research you have already conducted, you would likely find yourself in the trusted zone and will notice that the conversation does not suffer from skepticism from his or her end. Remember to compliment her or him about their career achievements and bring out the fact that you want to learn and grow similarly.

The key to identifying and nurturing a mentor-mentee relationship is the commitment from your side; this is not something that can work passively or as an afterthought. You need to be proactive, nurture the relationship, commit to a plan and bring your mentor along with you while you navigate that plan.

HOW CAN I ENHANCE MY RESUME?

The life cycle of the job search has become very complex. However, the curriculum vitae (CV) or resume continues to be one of the most vital steps, especially given the fact that if your resume fails to tell a compelling story upfront, the deal breaks even before you interact with a human.

Be mindful that if you create a resume that is not optimized to today's search engines and keyword filtering processes, your resume has a lower chance of being seen by the recruiter or hiring manager. As such, edit your resume to custom fit each job you apply for. Recruiters and hiring managers are also quick to catch cookie-cutter resumes, so tailoring is key.

What makes for a compelling and optimized resume that we speak of? Here are some ideas that make for a fit-for-purpose resume with the potential to pass the Artificial Intelligence test. Almost every organization utilizes automation and orchestration to filter through the thousands of resumes they receive and only forward the relevant ones to the hiring manager.

- Make it specific: The one-size-fits-all resume will not work. Period. You will need to ensure that your previous positions, skill set, education and background have the right keywords for the role you are targeting.

- Ideate but DO NOT plagiarize: It is okay to look for samples, templates and examples out there, but never plagiarize someone else's resume as you will fail when specific questions are asked. There is nothing wrong in using samples as a guide for ideas and formatting, but always take pride in building a unique storyline.

- BLUF (Bottom Line Up Front): Not the kind of "bluff" you were thinking. A BLUF is a very well-articulated summary as the first thing that is read on your cover letter or resume. The BLUF helps captivate the reviewer. The BLUF summary should ideally not be more than 200-250 words and briefly summarize what you are all about and what makes you someone the hiring manager wants to interview.

- Professional resume services: Remember, the only purpose of a resume is to get you an interview. A well-written and fit-for-purpose resume should get you just that. If after applying to many jobs and you still do not receive a call, then your resume is failing you. At this point, seek guidance from a professional resume writer. Some of us lack writing skills and/or might have language barriers if English is a second language. This is okay. It is important to research options, check reputation scores and customer feedback before paying for professional resume services. Also, be mindful of the timeline to establish a professional resume. If you feel you may need this service, it is best to add it to your career journey map early on so you are not up against a job posting deadline.

- It is true that, in general, most employers only care about receiving accurate information. In some cases, how it is presented professionally, what is highlighted and what is not highlighted in proper chronology do matter. A professional resume writer might be able to help with these. According to LinkedIn, an employer only spends about six seconds scanning a resume, and you really want to make the most of those six seconds.[43]

WHAT IF I DON'T GET A JOB INTERVIEW?

Now that you have a professional written and presentable resume, it's time to start your job search. Take the time to update your LinkedIn profile and your workplace intranet profile. As previously mentioned, leverage opportunities internally at your existing workplace first. If that is not an option, you are starting your job search as an external hire with a prospective company. Like with any job search of this nature, you must manage your expectations at the outset by understanding:

- there will be tough external competition;
- it may take some time depending on where you live and available positions applicable to your skill level (three months to one year);
- you may land a phone or video interview and not proceed to next-level interviews;
- you may proceed to next-level interviews with management, become shortlisted as a top candidate and not receive a job offer;
- there may be long gaps between interviews; or
- you do not receive an interview at all.
- Know you are not alone. According to Workopolis, only 2% of applicants get interviewed.[44] At first blush, that number can be discouraging. Your survival kit through the challenging job search phase is as follows:
- Don't put all your eggs in one basket. Apply to as many jobs you may be qualified for, even if you qualify only 60%. Do not underestimate yourself. If you believe you are closely qualified for the job, apply for it.

- Continuously edit your resume with job posting keywords applicable to you until you land a draft that starts to become noticed by recruiters and hiring managers.

- If you get even one interview, treat it as the only interview you will get. Preparation is key.

- If one interview is all you get, ask for feedback if they are willing to provide it. Then learn from it and apply. The key here is to fail fast, fail forward and not ruminate.

- Thank all your interviewers at every stage of the interview process.

- Continue to excel in your current job while continuing to network, attending relevant events and volunteering in the cybersecurity community, locally and online.

- If you are a cybersecurity undergraduate student or continuing education student who has excelled in courses, build relationships with your instructors. Many of them are acutely observing the talent pool in each cohort for potential future hires to work with them at their companies.

- Take time to pause from the job search if you experience fatigue or growing self-doubt. It's okay to take time out, recharge, do things you love, call someone from your Cyber5 Hotline, revive your self-esteem then restart when you've regained your optimism.

- Build job search emotional resilience: Common psychological injuries with any job search are rejection and failure. Therefore, establish boundaries with your job search; practice self-care of your mental, emotional and physical health; and bounce back when you are ready. Along the way, remind yourself why you started this journey, trust the process and be confident in your potential.

How do I prepare for a cybersecurity interview?

Congratulations on getting your resume noticed and shortlisted for an interview!

Like most interview processes, there is typically a phone interview first. If you do well on the phone interview, you can expect a series of in-person or virtual interviews with team members, the individual you would report to and possibly another senior individual.

To prepare for all these interviews, let's be honest – knowledge is power. Never go to an interview unprepared and without relevant knowledge about the role, the company and even the manager who interviews you. An instant personal connection can be made when you drive the conversation to common interests based on the research you have conducted.

Conducting proper research about the role you applied for and about the organization is critical. Research the company and the people on LinkedIn and see if you have any common connections. Then reach out to those common connections to see what can be learned about the company. Additionally, there are various applications that bring the news to you. No matter where you get your news from (Google News, Apple News, etc.), there is a way to personalize it to your needs.

Take the time to anticipate the interview questions relevant to the role you are being interviewed for and craft your responses. If you have access to LinkedIn Learning through school or your work, search job interview preparation videos that walk you through the common questions and pitfalls to avoid.

For intermediate and advanced-level roles, expect the interview questions to be tailored to specific responsibilities and technologies required of the role. For these roles, it is no longer applicable to simply state your experience generically, such as "10 years of risk management experience." You must elaborate on security risk management by stating your experience with conducting risk and maturity assessments using specific and relevant security frameworks and methodologies. Then you must be able to back that up with a particular project example, your role on the project, the problem, your contribution to solving the problem and the outcome. Sometimes, you will be asked a series of questions about technologies the role is required to use and state your level of experience on a scale of 1-10. If you state a high number out of 10, be prepared to speak to that specific technology and how you have used it and for when with specific examples.

Above all, be honest in your job interview. There is no need to embellish because hiring managers and recruiters can pinpoint dishonesty. If you are asked about how a technology works and do not know anything about it, simply state just that and add that you are familiar with perhaps a similar technology in your past experience.

Treat each interview as an opportunity to learn and grow. You may be pleasantly surprised to see how far you impress interviewers and hiring managers with your passion, past experience and authentic self.

How do I make myself stand out in a male-dominated field?

You are correct. Cybersecurity remains a male-dominated field. Now shift that known fact into opportunity. More and more companies are closing the gender gap through strategic hires and internal promotions. These companies are also purposefully rebuilding a culture of genuine inclusion where women and visible minorities have dedicated HR personnel internally to help them grow vertically or laterally.

Furthermore, there is an increase of resources available to working moms without hindrance to growth and development. While this will take more time to become fully realized, we are seeing the start of a global cultural shift in uplifting and supporting women and visible minorities as equals in the workforce.

Therefore, it is important for you—as a woman or as a visible minority woman in cybersecurity—to build your brand. Define what you want to be known for in the cybersecurity field and establish your story. You must be proactive in finding opportunities internal and external to your organization to showcase your passion, your expertise, your accomplishments and your humility. A story that resonates with many can create a substantial following.

Internally at your organization, you should proactively collaborate with your leaders and HR to determine what the future looks like as a female cybersecurity leader. Externally, you should be intentional in collaborating with your mentor(s), your network and your support system to help build your cybersecurity profile and increase visibility as a future cybersecurity leader to watch.

All it takes is for you to be brave, step out into the world and let the cybersecurity community know who you are. We will see you, hear you, support you and cheer you on.

WHAT IS A DAY-IN-THE-LIFE LIKE FOR A CYBERSECURITY PROFESSIONAL?

This depends on who you ask. Pre-COVID-19, a consulting manager may tell you she travels Monday to Thursday visiting and networking with clients. Behind the scenes, she's spending countless hours working on tight timelines to produce high-quality proposals and client deliverables.

Let's take, for example, a 20-year cybersecurity veteran who started as an analyst within the Security Operations Center (SOC) and now is a senior executive with a large consulting firm. For her, the trajectory has been fascinating, moving from a tightly process-driven day to now interacting with client executives, listening to their problems, advising them about risks and budgets and, in some cases, acting as a virtual-CISO for a client. The bottom line is there is no "cookie-cutter" cyber career progression.

If you work at a bank, your cyber career evolves differently as compared to someone who works at a utility company. At a bank, you are focused on protecting customer financial information and are closely associated with Payment Card Industry (PCI) standards. At a utility company, you work closely with the requirements of the NERC (North American Electric Reliability Corporation) CIP (Critical Infrastructure Protection) standard, which means you are required to protect critical infrastructure and operational technology. These career evolution examples are vastly different, and your career journey will be unique day after day, year after year.

How do I have a fruitful and long-term career in cybersecurity?

You are in the driver's seat when it comes to your career in cybersecurity. Change is scary. There's no doubt about that. We encourage you to get out of your comfort zone and believe in yourself.

Easier said than done. You are right there too.

Sometimes planning long-term creates heart palpitations, heartburn, stress eating and more. Thinking too big and too broad all at once can be stressful. To make career planning more manageable, take bite-size chunks and start there. You don't need to know your long-term journey now. You just need to focus on one or two things you need to achieve first- that's it. Once you've achieved those two things, add two more steps to your plan. Then continue to build and act on your plan with the same cadence over time.

Once you have a foot in the door as a practitioner, you will have the opportunity to learn and continually improve your soft skills as well as business and technical skills by way of the 70/20/10 rule-70% hands-on experience, 20% learn from others and 10% formal training.

Through projects and assignments, you will harness your strengths and determine which domain(s) you can truly excel as a technical specialist, non-technical specialist or management. It is also okay to change mentors over time as well. As you grow, you will determine who the right mentor is to complement your career journey.

As with any new career, there will be challenges along the way. Ensure to keep your Cyber5 Hotline on speed dial and keep your mentor apprised so they can provide guidance and support through difficult times. Don't forget to thank your supporters along the way.

When ready, give back through volunteer work and mentorship. There will be many more who will need your kindness, expertise and guidance to help them navigate their cybersecurity journey.

Above all, never stop being curious. Never stop learning. This will give you the edge among your competitors and help you stand out as you keep growing. And don't be afraid to dream big. You've come this far, haven't you?

Chapter 7 Key takeaways

As a cybersecurity job seeker:

- Determine what you want to be when you grow up in cybersecurity (technical, non-technical or managerial) by following the SIM method: select your specialization, interview a cybersecurity practitioner and map your journey.

- Obtain the necessary cybersecurity education, training and certification to help you stand out as a competitive candidate.

- Collaborate with mentors.

- Excel in your pursuits because your reputation travels with you no matter where you land.

- Prepare for the short-and-long haul of the job search by managing your expectations and building resilience.

- Get out of your comfort zone, action your plan and believe in yourself. When you believe in yourself, the world will too.

Conclusion

There are junctures in most professional careers where a choice needs to be made. Most choices that reap satisfying benefits later involve a substantial amount of prep work, learning new things (unlearning a few), meeting new people and pushing the envelope to the limit of your comfort zone. That said, you are your greatest investment and...

> *"The greatest project you'll ever work on is you." –* Sonny Franco

Investing in your future begins with defining what you want to be when you grow up and what you want to be known for. Regardless of the career you choose, there are various things you must achieve to obtain that career, including years of post-secondary education and work experience augmented with market-relevant certifications, networking with professional organizations, maintaining career and life coaches and dedicating time to your professional brand. These steps often lead to a well-rounded professional career.

Although nobody can predict the future, logical thinking and evidence-based actions can lead to somewhat predictable outcomes. One such logical way of thinking would be to assume a fast-paced "move to digital." Cybersecurity underpins this digital journey. Be it change to regulatory compliance or keeping ahead of the adversary, the need for appropriately skilled cybersecurity professionals is on-demand now more than ever.

Through the various chapters in this book, we shared field-tested perspectives to either start, specialize, super specialize and help increase the breadth of your existing experience within the cybersecurity profession. As educators, we were already answering numerous questions about the areas covered in this book, so putting it all together for you and everyone like you to benefit from made much sense. Although not all aspects covered in this book may apply to your unique situation, much of it is useful in cybersecurity and beyond.

As the battered global economy emerges from the fallout of the COVID-19 pandemic, the business landscape has evolved with resiliency, security and automation top of mind. The workplace culture is changed for good as employees have adapted to the new normal of work-from-home and virtual meetings. The world we live and work in today is vastly different from the world before the pandemic.

Indeed, be it due to socioeconomic, political or natural reasons, there are times when the status quo is no longer applicable, and some drastic steps are needed. These drastic steps sometimes translate to a career overhaul or at least a significant change in direction in a career progression. As Danielle LaPorte aptly inspired in a #truthbomb: "You can be scared and really ready."

A rewarding career in cybersecurity is at your fingertips. So, what are you waiting for?

Acknowledgments

Ali Khan

I would like to thank my parents, Wasi Khan and Ghazala Talat for building the foundations in me by leading and paving the way for me to get the best education in the world. I still remember my father's struggle in the early 2000s as we looked to make ends meet having recently immigrated to Canada, which is now our beloved home and by far, the best family decision ever taken.

I am also thankful for all the support my younger sister, Mariam Khan, my brother-in-law, Awais Rehman, my brother, Ahmad Khan, my sister-in-law, Hanna Ameen and their families have provided to bring this to fruition. To my grandmother, Jameela Khan, our family's head of the state, what an honor for me to continue to learn from you what patience and virtue is all about.

A special thanks to my wife, Shumaila Khan, who stands by me as a stronger reflection of myself. Your sacrifices and support to help me carry on my career and the peace that you bring within our lives are beyond human. To my angel daughters, Dina Khan, Amna Khan and Zaina Khan, thank you for supporting your father during all the days and nights sacrificed to bring this to the world. Your future success is the fuel that keeps me going.

Last but not the least, I thank you and those who will follow me and take from me what is good. You inspire me to do more; the pursuit of purpose in your eyes and our conversations fuels me to keep doing more, for you, for us. I dedicate this book to my father, Wasi Khan. #thecyberAli

Gaurav Kumar

This book came to be with many ifs and buts, and during one of the most testing times in my life; however, it quickly became one of the most rewarding endeavors I could have ever imagined. I express my deep gratitude to the people mentioned here for their advice, patience, and unrelenting faith in whatever craziness I have ever embarked on.

To the strongest woman I know, my wife, Shivani Kumar. This book came to conception while we were fighting the toughest battle of our lives: your cancer. Despite the uncertainties, you inspired me every day and always had happy and positive feelings around me. I am convinced that this would not have been without you.

To my son, Vedaant Kumar, who is the most admirable human being in my life. I apologize for taking away the time that belonged to you and thank you for never complaining. Thank you for keeping me honest and critiquing me when I needed it the most. You make me proud every day.

To my brother, Vivek Kumar, who continues to have more confidence in me than anyone else. Thanks for supporting me through every difficulty in my life and for being the flagbearer for our family's pride and joy.

To my parents, Rita and Satish Leal, for being the cornerstone of my support system and creating the foundation in me centered on curiosity. You taught me that by simply being friendly, respectful, and thoughtful of others, one could realize utmost internal satisfaction.

I am eternally grateful to all my students who made me realize the need for a book like this and who continue to push me towards getting better relentlessly.

I dedicate this book to my father-in-law, Surinder Sharma, who we recently lost to COVID-19, and whose presence is felt every moment. You taught me discipline, tough love, the power of planning, and, more importantly, to always keep my chin up. I am a better human being because of you.

Arlene Worsley

When I decided to transition into cybersecurity mid-career, I knew it would take a vast amount of courage, sacrifice and a village to see it through. As a single mom and a working professional, there were times when I doubted my decision and feared the consequence of failure. There was just so much more at stake with two kids to support on my own. It was on the shoulders of my family and dear friends where I asked, "What if I fail?" Because of their enthusiasm and encouragement, I learned to embrace the right question, "What if I succeed and more?" Now here we are celebrating what success and more can be like together.

To my mom, Evelyn Juan, I couldn't have done any of this without you. Thank you for taking care of the kids and me. To my dad, Randy McKilligan, thank you for always believing in me since day one. To my aunt, Maria Ochoa, thank you for being my champion. To my dear friend Tatiana Tomljanovic-Wilson, thank you for standing by me through it all. To Jeff Whiting, thank you for helping me unlock and unleash my true self through kindness and laughter. To my co-authors Ali Khan and Gaurav Kumar, thank you for taking me on this incredible journey during a time that tested our grit to see it through—we did it, and it is certainly worth every moment. And above all, to my kids Tony and Mia, you are my strength, inspiration and reason. I dedicate this book to you.

With sadness and in loving memory, my family lost Florencio Juan and Scott McKilligan just before publishing this book. We lost you too soon. Our fondest memories are those filled with your humor and laughter. When my kids and I look at the night sky, we will see you in the stars and carry the joy you brought to our lives forever in our hearts.

THE AUTHORS

Having been brought up in over three continents as part of his childhood, **Ali Khan, CCISO, CISM, CISSP, CDPSE, CISA** also known as *#thecyberAli*, is an executive cybersecurity and risk management professional with an "out-of-the-box" thinking approach, Ali serves the cybersecurity industry globally with his innovative approach to executing and delivering on cybersecurity initiatives. With strong attention to detail, Ali builds a high level of trust and strong relationships with his clients to effectively execute on the complex requirements cybersecurity initiatives carry. As a mentor and educator, Ali continues to work closely with educational institutions to develop the next generation of cybersecurity professionals and also provides mentorship, guidance and career coaching to upcoming and aspiring cyber professionals and startups. Ali also volunteers his time and efforts at not-for-profit organizations providing his subject matter expertise. Ali has an Executive Management Certificate in Leading Organizations in Disruptive Times from INSEAD School of Business, France, an Honors Bachelor of Arts, Majoring in Information Technology Management from York University, Canada and is a Certified Chief Information Security Officer (CCISO), Certified Information Security Manager (CISM), Certified Information Systems Security Professional (CISSP), Certified Data Privacy Solutions Engineer (CDPSE) and Certified Information Systems Auditor (CISA). Ali also has NATO and Level II (Secret) Clearance from the Government of Canada. Ali is married to his beautiful wife, Shumaila and is blessed with three angel daughters, Dina, Amna and Zaina who are his gifts of this life. Ali loves cars and likes to participate in a number of sports ranging from cricket, soccer and basketball. For more details, visit his website at www.thecyberali.com and follow him on online on social media platforms using hashtag *#thecyberAli.*

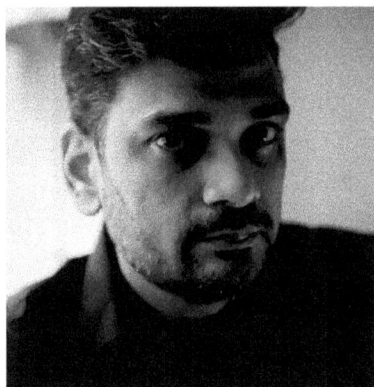

Gaurav Kumar, CISSP, CISM, CRISC, CDPSE, CCSK, SABSA (SCF), is currently a senior cybersecurity and privacy consulting executive with one of the world's largest management consulting firms. He is focused on building cybersecurity strategies resulting in business-aligned security programs. Gaurav's experience spans almost two decades. He has worked in cybersecurity operations, risk management, compliance and governance, and assurance roles—his expertise spans across the telecommunication, retail, high-tech, finance, and health and public safety industries. Gaurav is also a passionate educator. He is currently an instructor and academic advisor at a top-ranked international teaching and research university, based in Toronto, Canada, where he contributed to the development of the Cybersecurity Program. He is also an avid tennis player and enjoys the outdoors. Gaurav has a master's degree in information security and assurance from Norwich University in Vermont, USA. Gaurav lives with his family in Toronto. To explore more about Gaurav and to stay connected, visit him on his website https://gkumar.ca.

Arlene Worsley, CISM, didn't follow the "normal" path into cybersecurity. She mastered her craft in reputation management and IT web operations leading organizations in navigating crises, digital disruption and transformation. She spent countless hours as a crisis management lead in incident command centers wearing a neon vest through workplace fatalities, natural disasters and evacuations, environmental incidents, anti-industry campaigns, physical protests, provincial court hearings, sophisticated cyberattacks resulting in data breaches and online brand abuse.

In late 2018, Arlene took a leap of faith and pursued, with fervent passion, a mid-career reinvention to specialize in cybersecurity. Today, Arlene is an award-winning cyber risk and crisis management expert, national cybersecurity assistant instructor, guest speaker, mentor and author.

Home for Arlene, a single mom of two kids, is in Calgary, Canada, where she is an experienced all-season hiker and explorer of the majestic Rocky Mountains achieving one summit after another. Roughing the scree at steep terrain, repeatedly falling down, getting back up and facing her intense fear of heights is symbolic of how she has overcome trauma and adversity with courage, tenacity and heart. For more about her brave story, visit www.cyberandsapphire.com.

ENDNOTES

[1] *(ISC)² Report Finds Cybersecurity Workforce Gap Has Increased to More Than 2.9 Million Globally.*
www.isc2.org/News-and-Events/Press-Room/Posts/2018/10/17/ISC2-Report-Finds-Cybersecurity-Workforce-Gap-Has-Increased-to-More-Than-2-9-Million-Globally

[2] NeSmith, Brian. *Council Post: The Cybersecurity Talent Gap is an Industry Crisis.* Forbes Magazine, 13 Aug. 2018.
www.forbes.com/sites/forbestechcouncil/2018/08/09/the-cybersecurity-talent-gap-is-an-industry-crisis/#1a9923afa6b3

[3] *(ISC)² Research Finds Women Comprise 24% of Global Cybersecurity Workforce*, 2 Apr. 2019. www.isc2.org/News-and-Events/Press-Room/Posts/2019/04/02/ISC2-Research-Finds-Women-Comprise-24-percent-of-Global-Cybersecurity-Workforce

[4] *Staying cyber-healthy during COVID-19 isolation.* Canadian Centre for Cyber Security, 09 Apr. 2020.
cyber.gc.ca/en/news/staying-cyber-healthy-during-covid-19-isolation

[5] Ranjan, Ruhi. *Accenture: Navigating the Human and Business Impact of COVID-19.* Ethisphere Magazine, 2020.
magazine.ethisphere.com/bela-south-asia-accenture

[6] Marr, Bernard. *How Much Data Do We Create Every Day? The Mind-Blowing Stats Everyone Should Read.* Forbes Magazine, 5 Sept. 2019.
www.forbes.com/sites/bernardmarr/2018/05/21/how-much-data-do-we-create-every-day-the-mind-blowing-stats-everyone-should-read/#3901bc0760ba

[7] *Creeper (Program)*. Wikipedia, Wikimedia Foundation, 20 Mar. 2021. en.wikipedia.org/wiki/Creeper_(program)

[8] *(ISC)2 Member Counts*. 01 Jan. 2021. www.isc2.org/About/Member-Counts

[9] Fox, Chris. *Google Hit with £44m GDPR Fine over Ads*. BBC News, BBC, 21 Jan. 2019. www.bbc.com/news/technology-46944696

[10] (ISC)². *Code of Ethics: Complaint Procedures: Committee Members*. Code of Ethics | Complaint Procedures | Committee Members. www.isc2.org/Ethics

[11] *SolarWinds Security Advisory*. SolarWinds, 2021. www.solarwinds.com/sa-overview/securityadvisory

[12] PayScale. *Average Chief Information Security Officer Salary*. www.payscale.com/research/US/Job=Chief_Information_Security_Officer/Salary

[13] ZipRecruiter. *Information Security Architect Salary*. www.ziprecruiter.com/Salaries/Information-Security-Architect-Salary

[14] PayScale. *Average Information Security Manager Salary*. www.payscale.com/research/US/Job=Information_Security_Manager/Salary

[15] *Project Management Professional (PMP)*. Project Management Institute. www.pmi.org/certifications/project-management-pmp

[16] ZipRecruiter. *IT Security Project Manager Salary*. www.ziprecruiter.com/Salaries/IT-Security-Project-Manager-Salary

[17] PayScale. Average Security Engineer Salary. www.payscale.com/research/US/Job=Security_Engineer/Salary

[18] PayScale. *Average Security Incident Response Salary.* www.payscale.com/research/US/Job=Security_Incident_Response/Salary

[19] PayScale. *Average Penetration Tester Salary.* www.payscale.com/research/US/Job=Penetration_Tester/Salary

[20] PayScale. *Average Forensic Computer Analyst Salary.* www.payscale.com/research/US/Job=Forensic_Computer_Analyst/Salary

[21] PayScale. *Average Malware Analyst Salary.* www.payscale.com/research/US/Job=Malware_Analyst/Salary

[22] PayScale. *Average Cyber Security Analyst Salary.* www.payscale.com/research/US/Job=Cyber_Security_Analyst/Salary

[23] U.S. News. *Information Security Analyst Salary.* Best Technology Jobs. money.usnews.com/careers/best-jobs/information-security-analyst/salary

[24] PayScale. *Average Security Engineer, Information Systems Salary.* www.payscale.com/research/US/Job=Security_Engineer%2C_Information_Systems/Salary

[25] PayScale. *Average Cyber Security Analyst with Security Risk Management Skills Salary.* www.payscale.com/research/US/Job=Cyber_Security_Analyst/Salary/bb8295a5/Security-Risk-Management

[26] Indeed.com. *How much does a Legal Advisor make in the United States?* www.indeed.com/salaries/legal-advisor-Salaries

[27] PayScale. *Average Privacy Officer Salary.* www.payscale.com/research/US/Job=Privacy_Officer/Salary

[28] ZipRecruiter. *Cyber Security Sales Salary.* https://www.ziprecruiter.com/Salaries/Cyber-Security-Sales-Salary

[29] Glassdoor. Cyber Security Researcher Salaries. www.glassdoor.com/Salaries/cyber-security-researcher-salary-SRCH_KO0,25.htm

[30] PayScale. *Average Security Assessor Salary.* https://www.payscale.com/research/US/Job=Security_Assessor/Salary

[31] ZipRecruiter. *Cyber Security Auditor Salary.* www.ziprecruiter.com/Salaries/Cyber-Security-Auditor-Salary

[32] PayScale. *Average Security Consultant, IT Salary.* www.payscale.com/research/US/Job=Security_Consultant%2C_IT/Salary

[33] ZipRecruiter. *Cyber Security Instructor Salary.* www.ziprecruiter.com/Salaries/Cyber-Security-Instructor-Salary

[34] *What Are the Best Cyber Security Certifications in 2021? (List of the Top 10).* InfoSec Careers Network, 10 Mar. 2021. www.infosec-careers.com/the-best-cyber-security-certifications-in-2020

[35] RapidScale Follow. *Cloud Computing Stats - Security and Recovery.* SlideShare, 30 Jan. 2015. www.slideshare.net/rapidscale/cloud-computing-stats-security-and-recovery

[36] Jahncke, Stephen and Isabelle Lee. *It's time: Organize your future with Robotic Process Automation.* 28 Oct. 2016. PWC Canada. https://www.pwc.com/us/en/industry/entertainment-media/publications/tmm/assets/tmm-supply-chain.pdf

[37] Bailey, Stephanie. *Drones could help fight coronavirus by air-dropping medical supplies.* CTV News – CNN Digital. 31 Mar 2021. www.ctvnews.ca/sci-tech/drones-could-help-

fight-coronavirus-by-air-dropping-medical-supplies-1.5370312

[38] Townsend, Kevin. *Sky-high concerns: Understanding the security threat posed by drones.* Avast.com. 26 Sept 2019. blog.avast.com/what-security-threats-are-posed-by-drones

[39] *3D Printing Growth Accelerates Again.* Deloitte Insights, www2.deloitte.com/us/en/insights/industry/technology/technology-media-and-telecom-predictions/3d-printing-market.html.

[40] *2019 Global Threat Report.* Crowdstrike.com, 16 Apr. 2020. www.crowdstrike.com/resources/crowdcasts/2019-global-threat-report-crowdcast

[41] *Ponemon Study on Gaps in Vulnerability Response.* ServiceNow. www.servicenow.co.uk/lpayr/ponemon-vulnerability-survey.html

[42] Glassdoor Team. *5 More Companies That No Longer Require a Degree—Apply Now.* Glassdoor, 01 Jan 2020. www.glassdoor.com/blog/no-degree-required

[43] Friedman, Andrew. *6 Seconds is the Average Time Spent Reading a Resume.* 16 Feb 2017. www.linkedin.com/pulse/six-seconds-average-time-spent-reading-resume-andrew-j-friedman

[44] *Why only 2% of applicants actually get interviews.* Workopolis, 10 Nov 2016. careers.workopolis.com/advice/only-2-of-applicants-actually-get-interviews-heres-how-to-be-one-of-them

www.ingramcontent.com/pod-product-compliance
Lightning Source LLC
Chambersburg PA
CBHW070922210326
41520CB00021B/6770